T0205406

# Intelligent Systems Reference Library

Volume 127

*About this Series*

The aim of this series is to publish a Reference Library, including novel advances and developments in all aspects of Intelligent Systems in an easily accessible and well structured form. The series includes reference works, handbooks, compendia, textbooks, well-structured monographs, dictionaries, and encyclopedias. It contains well integrated knowledge and current information in the field of Intelligent Systems. The series covers the theory, applications, and design methods of Intelligent Systems. Virtually all disciplines such as engineering, computer science, avionics, business, e-commerce, environment, healthcare, physics and life science are included.

More information about this series at http://www.springer.com/series/8578

Amit Konar · Diptendu Bhattacharya

# Time-Series Prediction and Applications

## A Machine Intelligence Approach

Amit Konar
Artificial Intelligence Laboratory, ETCE Department
Jadavpur University
Kolkata
India

Diptendu Bhattacharya
Department of Computer Science and Engineering
National Institute of Technology, Agartala
Agartala, Tripura
India

ISSN 1868-4394 ISSN 1868-4408 (electronic)
Intelligent Systems Reference Library
ISBN 978-3-319-85435-9 ISBN 978-3-319-54597-4 (eBook)
DOI 10.1007/978-3-319-54597-4



Printed on acid-free paper

This Springer imprint is published by Springer Nature
The registered company is Springer International Publishing AG
The registered company address is: Gewerbestrasse 11, 6330 Cham, Switzerland

*Dedicated to*
***Shirdi Sai Baba***

# Preface

A time-series describes a discrete sequence of amplitudes of an entity ordered over time. Typically, a time-series may describe the temporal variations in atmospheric temperature, rainfall, humidity, stock price, and any other measurable quantity that has a variation with time and available as a discrete data points sampled uniformly or non-uniformly over time. Prediction of a time-series refers to determining its value at unknown time point $t + 1$ from the knowledge of the time-series at its current time t and also previous time points $(t - 1)$, $(t - 2)$, …, $(t - n - 1)$. On occasions, the predicted value of the time-series depends also on one or more influencing time-series, besides having dependence on its past values. A time-series prediction thus refers to a regression problem in high-dimensional space, where the predicted value describes a highly nonlinear function of its past values and also other relevant time-series. Unfortunately, the nature of nonlinearity being unknown adds more complexity to the prediction problem.

Several approaches to time-series prediction are available in the literature. One of the early works on time-series prediction refers to thc well-known ARMA (autoregressive moving average) model, which is appropriate for prediction of a stationary time-series. However, many of the natural/man-made time-series are non-stationary with varying frequency components over the time frames. One indirect approach to predict a non-stationary time-series is to transform it to an equivalent stationary series by taking the differences of consecutive data points and then to test its stationarity with the help of autocorrelation (AC) and partial autocorrelation (PAC) measures. In case the resulting time-series is not stationary yet, we repeat the above step until stationarity is attained. The prediction of the resulting stationary time-series is performed using the ARMA model, and the predicted series is integrated as many times the difference operator has been applied to the original time-series. The complete process of predicting a non-stationary time-series by adoption of the above three steps, transformation to a stationary time-series, prediction using the ARMA model, and integration of the predicted stationary time-series by the requisite number of times is referred to as ARIMA (autoregressive integrated moving average) model.

The fundamental limitation of the ARIMA model lies in the high-order differentiation of the series, which sometimes results in a white noise with zero mean, and therefore, it is of no use from the prediction point of view. This calls for alternative formulation to handle the problem. The logic of fuzzy sets has an inherent advantage to represent nonlinear mapping, irrespective of non-stationary characteristics of the time-series. In addition, the non-Gaussian behavior of a time-series can be approximated by a locally Gaussian formulation. Thus, fuzzy logic can handle nonlinearity, non-stationarity, and non-Gaussian-ness of the time-series. In addition, non-deterministic transition of the time-series from a given partition of a vertically partitioned time-series to others can also be handled with fuzzy logic by concurrent firing of multiple rules and taking aggregation of the resulting inferences generated thereof. Besides fuzzy logic, pattern clustering and classification by neural and/or other means can also take care of the four issues indicated above. This motivated us to develop new prediction algorithms of time-series using pattern recognition/fuzzy reasoning techniques.

Song and Chissom in 1994 proposed fuzzy time-series, where they assign a membership value to each data point to represent the degree of belongingness of each data point in a given partition. They extracted single-point prediction rules from the time-series, where each rule represents a partition $P_i$ in the antecedent and a partition $P_j$ in the consequent, where partition $P_i$ includes the current time point and $P_j$ the next time point. Once the construction of rules is over, they developed fuzzy implication relations for each rule. These relations are combined into a single relational matrix, which is used later to derive fuzzy inferences from membership functions of measured time-series value at the current time point. A defuzzification algorithm is required to retrieve the predicted sample value from the fuzzy inference.

Extensive works on fuzzy time-series have been undertaken over the last two decades to perform prediction from raw time-series data, primarily to handle uncertainty in diverse ways. A few of these that need special mention includes partition width selection, influence of secondary data for the prediction of main time-series, extension of fuzzy reasoning mechanism (such as many-to-one mapping), different strategies for membership function selection, and the like. Pedrycz et al. introduced a novel technique to determine partition width in the settings of optimization and employed evolutionary algorithm to solve the problem. Chen et al. proposed several interesting strategies to utilize secondary memberships for the prediction of main factored economic time-series. They used more stable Dow Jones and NASDAQ as the secondary factor time-series for the TAIEX time-series as the main factor. Huarang et al. proposed different type-1 fuzzy inferential schemes for prediction of time-series. The details of the above references are given in this book.

In early studies of time-series prediction, researchers took active interest to utilize the nonlinear regression and functional approximation characteristics of artificial neural networks to predict time-series from their exemplary instances. Traces of research considering supervised learning techniques for prediction of time-series value are available in the literature. Early attempts include developing functional mapping for predicting the next time point in the series from the current time point value. Most of the researchers employed gradient descent learning-based back-propagation algorithm neural technique and its variant for time-series prediction. Among other significant works, neural approaches based on the radial basis function and support vector machine need special mention for time-series prediction. In recent times, researchers take active interest on deep learning and extreme learning techniques for time-series prediction.

This book includes six chapters. Chapter 1 provides a thorough introduction to the problem undertaken with justification to the importance of the selected problem, limitations of the existing approaches to handle the problem, and the new approaches to be adopted. Chapters 2–5 are the original contributions of the present research. Chapter 2 examines the scope of uncertainty modeling in time-series using interval type-2 fuzzy sets. Chapter 3 is an extension of Chap. 2 with an aim to reason with both type-1 and type-2 fuzzy sets for the prediction of close price time-series. The importance of using both type-1 and type-2 fuzzy sets is apparent from the availability of number of data points in a given partition of a prepartitioned time-series. When there exist a fewer data points in a single contiguous region of a partition, we represent the partition by a type-1 fuzzy set, else we go for an interval type-2 representation. Chapter 4 introduces a clustering technique for subsequence (comprising a few contiguous data points) prediction in a time-series. This has a great merit in forecasting applications, particularly for economic time-series. Chapter 5 attempts to design a new neural technique to concurrently fire a set of prediction rules realized on different neural networks and to combine the output of the neural networks together for prediction. Chapter 6 is the concluding chapter, covering the self-review and future research directions.

In Chap. 2, a new technique for uncertainty management in time-series modeling is proposed using interval type-2 fuzzy sets. Here, the time-series is partitioned into k number of blocks of uniform width, and each partition is modeled by an interval type-2 fuzzy set. Thus, transition of consecutive data points may be described by type-2 fuzzy rules with antecedent and consequent containing partitions $P_i$ and $P_j$, respectively, where $P_i$ and $P_j$ denote two consecutive data points of the time-series. We then employ interval type-2 fuzzy reasoning for prediction of the time-series. The most important aspect of the chapter is the inclusion of secondary factor in time-series prediction. The secondary factor is used here to indirectly control the growth/decay rate in the main factored time-series.

Chapter 3 is an extension of Chap. 2 to deal with fuzzy reasoning using both type-1 and interval type-2 fuzzy sets. The motivation of using both types of fuzzy sets has been discussed, and principles used for reasoning with prediction rules containing both types of fuzzy sets are discussed. The improvement in performance of the proposed technique with only type-1 and interval type-2 has been demonstrated, indicating the significance of the technique.

Chapter 4 deals with a novel machine learning approach for subsequence (comprising a few contiguous data points) prediction. This is undertaken in three phases, called segmentation, clustering, and automaton-based prediction. Segmentation is required to represent a given time-series as a sequence of segments (time blocks), where each segment representing a set of contiguous data points of any arbitrary dimension should have a semantically meaningful shape in the context of an economic time-series. In the current book, segmentation is performed by identifying the slopes of the data points in the series for a sequence of a finite number of data points and demarcating them into positive, negative, or near-zero slopes.

A segment is labeled as rising, falling, or having a zero slope, depending on the maximum frequency count of any of the three primitives: positive, negative, or near-zero slope, respectively, for every five or more consecutive data points in the series. After the segments are identified, we group them using a clustering algorithm based on shape similarity of the normalized segments. As the number of clusters is an unknown parameter, we used the well-known DBSCAN algorithm that does not require the said parameter. In addition, we extend the DBSCAN clustering algorithm to hierarchically cluster the data points based on descending order of data density. The cluster centers here represent structures of specific geometric shape describing transitions from one partition to the other. To keep track of the transition of one partition to the others using acquired structures, we develop an automaton and use it in the prediction phase to predict the structure emanating from a given time point. The predicted information includes the following: Given a terminating partition, whether there exists any feasible structure that terminates at the desired partition starting at the given time point? In addition, the probable time to reach the user-defined partition and its probability of occurrence are additional parameters supplied to the user after prediction. Experiments undertaken on three standard time-series confirm that the average accuracy of structure prediction is around 76%.

In Chap. 5, we group prediction rules extracted from a given time-series in a manner, such that all the rules containing the same partition in the antecedent can fire concurrently. This requires realization of the concurrently fireable rules on different neural networks, pretrained with supervised learning algorithms. Any traditional supervised learning algorithms could be used to solve the problem. We, however, used the well-known back-propagation algorithm for the training of the neural networks containing the prediction rules.

This book ends with a brief discussion on self-review of this book and relevant future research directions in Chap. 6.

Kolkata, India

Amit Konar
Diptendu Bhattacharya

# Acknowledgements

Authors gratefully acknowledge the support they received from individuals to write this book in its current form. Although the authors cite the contributions of a few individuals, they are grateful to all people who have helped in whatever way possible for the development of this book.

The first author gratefully acknowledges the support he received from his colleagues, friends, and students, without whose help and support this book could not be prepared in its current form. The second author equally is grateful to all his colleagues of National Institute of Technology (NIT), Agartala, whose unfailing help and assistance nurtured the progress of this book. The authors would also like to thank Prof. Suranjan Das, vice-chancellor of Jadavpur University, and Prof. Gopal Mugeraya, Director, NIT, Agartala, for providing them good and resourceful environments to develop this book. The authors are equally indebted to Prof. P. Venkateswaran, HOD, Department of Electronics and Telecommunication Engineering, Jadavpur University, for providing all sorts of support the authors required to complete this book in the present form.

Authors are grateful to the entire Springer for their hard work and kind support at different levels of production of this book. They also like to thank Prof. Lakhmi C. Jain whose active collaboration and support enriched this book in different forms.

Lastly, the authors acknowledge the support they received from their family members. In this regard, the first author recognizes the support of his wife: Srilekha, son: Dipanjan, and mother: Mrs. Minati Konar. The second author also acknowledges the support he received from his wife: Papri, son: Diptaraj, daughter: Indushree, and father: Shri Dhirendra Chandra Bhattacharya.

Amit Konar
Diptendu Bhattacharya

# Contents

# About the Authors

**Amit Konar** (SM'10) received the B.E. degree from Indian Institute of Engineering Science and Technology (formerly Bengal Engineering College), Howrah, India, in 1983, and the M.E., M.Phil., and Ph.D. degrees from Jadavpur University, Kolkata, India, in 1985, 1988, and 1994, respectively.

He is currently a Professor with the Department of Electronics and Telecommunication Engineering, Jadavpur University, where he is also the founding Coordinator of the M.Tech. Program on Intelligent Automation and Robotics. He has supervised 20 Ph.D. theses. He has over 300 publications in international journals and conference proceedings. He is the author of eight books, including two popular textbooks, *Artificial Intelligence and Soft Computing* (Boca Raton, FL, USA: CRC Press, 2000) and *Computational Intelligence: Principles, Techniques and Applications* (New York, NY, USA: Springer, 2005).

Dr. Konar is currently serving as an Associate Editor of IEEE Trans. on Fuzzy Systems, IEEE Trans. on Emerging Topics in Computational Intelligence, and *Neurocomputing, Elsevier*. He also served as an Associate Editor of IEEE Trans. on Systems, Man and Cybernetics: Part-A and Applied Soft Computing, Elsevier. He is the Editor-in-Chief of a Springer Book Series on Cognitive Intelligence and Robotics.

**Diptendu Bhattacharya** received the B.E. degree from National Institute of Technology (formerly Malaviya Regional Engineering College), Jaipur, Rajasthan, India, and M.E. Tel.E. (Computer Engineering) from Jadavpur University, Kolkata, India, in 1988 and 1999, respectively. He completed Ph.D. (engineering) as a QIP Ph.D. (engineering) fellow under the supervision of Prof. Amit Konar in Jadavpur University, Kolkata, in 2016. He is currently an Associate Professor in the Department of Computer Science and Engineering, National Institute of Technology, Agartala, Tripura, India. He stood first class first position in order of merit in his M.E. Tel.E. Program. Dr. Bhattacharya was the Head of the Department of Computer Science and Engineering, in National Institute of Technology, Agartala. He was also the winner of Weekly Nifty Prediction Contest organized by Personal Wealth Management Solutions Pvt. Ltd., Kolkata, India, in 2013. He is popular among his students. Also, Dr. Bhattacharya supervised several B.Tech. and M.Tech. theses in his teaching period in NIT, Agartala, India. His current research interests include type-2 fuzzy sets, artificial intelligence, computational intelligence, fuzzy time-series, and its prediction.

# Chapter 1
# An Introduction to Time-Series Prediction

**Abstract** This chapter provides an introduction to time-series prediction. It begins with a formal definition of time-series and gradually explores possible hindrances in predicting a time-series. These hindrances add uncertainty in time-series prediction. To cope up with uncertainty management, the chapter examines the scope of fuzzy sets and logic in the prediction of time-series. Besides dealing with uncertainty, the other important aspect in time-series prediction is to learn the structures embedded in the time-series. The chapter addresses the scope of machine learning in both prediction of the series and also the structures hiding inside the series. The influence of secondary factors in the main-factor time-series is reviewed and possible strategies to utilize secondary factors in predicting main factor time-series are addressed. The methodologies used to partition the dynamic range of a time-series for possible labeling of the diurnal series value in terms of partition number and also for prediction of the next time-point value in terms of the partition number are reviewed, and possible strategies for alternative approaches to partitioning the time-series are overviewed. The chapter ends with a discussion on the scope of the work, highlighting the goals and possible explorations and challenges of economic time-series prediction.

## 1.1 Defining Time-Series

A time-series is a discrete sequence of a time-valued function ordered over time. There exist a lot many real-world entities, such as rainfall, atmospheric temperature, population growth, gross domestic product (GDP) and the like, where the data are measured at a regular interval of (real) time. These unprocessed data, representing the temporal variations of the entity at fixed time-points, within a given finite interval of time together describes a time-series. The inclusion of finite interval of real time in the definition of a time-series makes sense from the perspective of its prediction at a time point not included in the recorded data set. However, if the motivation of the series is to preserve historical data only, ignoring futuristic



A. Konar and D. Bhattacharya, *Time-Series Prediction and Applications*,
Intelligent Systems Reference Library 127,
DOI 10.1007/978-3-319-54597-4_1

predictions, the restriction on fixed and finite interval from the notion of time-series can be dropped. For example, any discrete dynamical process response can be called a time-series, where the motivation is to examine the nature of dynamic behavior from the mathematical model of the series dynamics, does also represent a time-series.

Mathematicians represent a discrete process dynamics by a difference equation, where the solution of the difference equation describes a sequence of infinite series of time-valued function in an increasing order of time. Let f(t) be a continuous function over time t in the interval $[t_{min}, t_{max}]$. We discretize f(t) by f(kT), where T is the sampling interval and k = 0, 1, 2, 3, 4, … where $t_{min} = 0$ and $t_{max} = nT$, for some positive integer n. Given f(kT) for k = 0, 1, 2, …, n, time series prediction requires determining f(nT) using suitable techniques such as regression, heuristics, probabilistic techniques, neural net based supervised learning and many others.

## 1.2 Importance of Time-Series Prediction

Prediction of a time-series is useful for its wide-spread real-world applications. For example, the population growth in a country from the measure of its current population is useful to determine the future prospect of the citizens. In order to maintain a healthy environment of the citizens, the planning of proper utilization of national resources is required to channelize funds in diverse sectors as felt appropriate by the Government.

Secondly, the prediction of rainfall from the historical time-series data of previous years in a given locality of a country is an important issue to maintain necessary seasonal distribution of water for agricultural purposes.

Thirdly, the economic growth of a bank is an interesting time-series, controlled by several factors including, interest rate, loan sanctioned in the current year, population of new depositors and many others. The control policies of the bank on re-allocation of assets thus are detrimental to the forecasted time-series of its economic growth.

Enrollment of students in a department of a university also is a useful time-series for self-assessment of the department and the university as well in its international ranking, facilities offered to students and the job opportunities of the students. A fall-off in enrollment thus might be due to the failure to offer necessary facilities and infrastructures to students, which in turn determines the national and international ranking of the university.

Various financial institutions of the country have their investments in equities, derivatives, forex, commodities etc. They have also to maintain hedge funds. For all these kind of trading/investment activities a complete speculation is necessary. It is only possible with time series prediction of above trading/investment instruments (indices, commodities, forex, futures and options etc.

Thus wrong prediction of time series of instruments invites big losses for institutions, which may lead to bankruptcy. The correct prediction results, such as good profit in trading/investment, are always appreciated.

## 1.3 Hindrances in Economic Time-Series Prediction

A time-series used to describe stock index, generally, is non-linear, non-deterministic, non-Gaussian, and non-stationary. Non-linearity refers to non-linear variation of a discrete function with time. An economic time-series usually is non-linear. In general, there exists no straight-forward technique to determine the exact nature of the non-linearity of an economic time-series. This acts as hindrances to predict the time-series.

Certain real-world signals, such as electrocardiogram (ECG), maintain a periodic wave-shape. Prediction of such signals is relatively easier as the signals are predictive to some extent for their fixed wave-shape. These signals are often referred to as deterministic. An economic time-series like electroencephalogram (EEG) [1, 2] is a non-deterministic signal as it has no fixed wave-shape. Prediction of a non-deterministic signal is hard as the signal may take up any possible value within a wide range of lower and upper bounds, which too may vary under special situations. The methods adopted for the prediction of deterministic signal are not applicable for prediction of non-deterministic signals.

Most of the man-made signals used in laboratories for analog signal processing essentially contain fixed frequencies. Even in human-speech we have fixed formants, i.e., signal frequencies with a fundamental and several harmonics, but these frequencies do not change appreciably over time. Unlike the above situation, an economic time-series contains several frequencies that too vary widely over a finite time-interval of the signal. This characteristic of signals is referred to as non-stationarity. An economic time-series generally is non-deterministic, and thus requires special attention for its prediction.

Lastly, most of the commonly used signals support the Gaussian characteristics, i.e., the instantaneous values of the signal always lie within fixed bounds given by [mean $-$ 3 $\times$ SD, mean + 3 $\times$ SD], where mean and SD denotes the mean and standard deviation of the given signal. The Gaussian property follows from one fundamental observation that a 99% of the area under a Gaussian (bell-shaped) function lies in the above interval around the signal mean. Unfortunately, because of random fluctuation of an economic time-series, the signal sweeps widely and occasionally falls outside the given bounds and thus fails to satisfy the Gaussian characteristics of the signal. Consequently, an economic time-series is a non-Gaussian signal.

The non-stationary and non-Gaussian characteristics of a time-series act as fundamental hindrances in its prediction. In fact, because of these two characteristics, a time-series looks like a random fluctuation of the signal and thus requires non-conventional approaches for its prediction. In other words, the randomness in the time-series makes the prediction hard as it needs to deal with uncertainty management in the process of prediction.

Several approaches to time-series prediction are available in the literature. One of the early works on time-series prediction refers to the well-known ARMA (Auto Regressive Moving Average) model, which is appropriate for prediction of a stationary time-series. However, many of the natural/man-made time-series are non-stationary with varying frequency components over the time frames. One indirect approach to predict a non-stationary time-series is to transform it to an equivalent stationary series by taking the differences of consecutive data points, and then to test its stationarity with the help of auto-correlation (AC) and partial auto-correlation (PAC) measures. In case the resulting time-series is not stationary yet, we repeat the above step until stationarity is attained [3–7]. The prediction of the resulting stationary time-series is performed using the ARMA model and the predicted series is integrated as many times the difference operator has been applied to the original time-series. The complete process of predicting a non-stationary time-series by adoption of the above three steps: transformation to a stationary time-series, prediction using the ARMA model and integration of the predicted stationary time-series by the requisite number of times is referred to as ARIMA (Auto Regressive Integrated Moving Average) model [8–15].

The fundamental limitation of the ARIMA model lies in the high order differentiation of the series, which sometimes result in a white noise with zero mean, and therefore is of no use from the prediction point of view. This calls for alternative formulation to handle the problem. Fuzzy sets and neural networks have inherent advantage to represent non-linear mapping, irrespective of non-stationary characteristics of the time-series. In addition, the non-Gaussian behavior of a time-series can be approximated by a locally Gaussian formulation. Thus fuzzy logic can handle non-linearity, non-stationarity and non-Gaussian-ness of the time-series. In addition, non-deterministic transition of the time-series from a given partition of a vertically partitioned time-series to others can also be handled with fuzzy logic by concurrent firing of multiple rules and taking aggregation of the resulting inferences generated thereof. Besides fuzzy logic, pattern clustering and classification by neural and/or other means can also take care of the above four issues indicated above. This motivated us to develop new prediction algorithms of time-series using pattern recognition/fuzzy reasoning techniques.

## 1.4 Machine Learning Approach to Time-Series Prediction

Learning refers to natural acquisition of knowledge reflected by certain parametric changes in a system. It is synonymously used with pattern classification and clustering. However, in general, learning has much more scope than those of pattern classification/clustering. In a pattern recognition problem, we adapt system parameters to represent the measured input-output relationship in a system. This is often performed by a technique, popularly known as *supervised learning*. The name: supervised learning stems from the concept that such learning requires a set of training/exemplary instances (input and output combinations), generated by the trainer/supervisor prior to learning. After the learning (adaptation) phase is over, we use the adapted system parameters to predict system outputs for each known input. In a pattern clustering problem, we group objects in a cluster based on similarity in attributes of one objects with other objects. Clustering is advantageous when we do not have training instances for individual object class. Clustering is often referred to as unsupervised learning as it works without training instances.

Besides clustering and classification, there exist two other popular varieties of learning, well-known as reinforcement learning and competitive learning. In reinforcement learning, a system acquires knowledge based on reward and penalty mechanism. In other words, here an agent (a software/hardwired device) acts on its environment and a critic sitting in the environment measures the reward/penalty the agent should receive based on the effectiveness of the action in the environment with respect to a fixed goal. The agent records the reward and penalty during the learning phase, and uses the measure of reward/penalty to select the action at a given environmental state (situation) to yield its best response. The last kind of learning, called competitive learning, allows competition between two or more learning strategies to select the best in a given situation.

In this book, we would use both supervised and unsupervised learning for predictive applications. Unsupervised clustering can be employed on the similar sequence of data points representing structurally similar segments of a time-series. Such similarity helps in predicting a similar sequence from incomplete early sub-sequence in a time-series. For instance, suppose, we discovered 6 different structures comprising a fixed number of consecutive data points in a time-series. Let us call them sequence 1 through sequence 6. Now, we need to predict which particular sequence is expected to follow from the time-series points of last few days (say one week). This requires determining similarity of the daily time-series of last one week with known sequences 1 through 6, and supposes sequence 4 is the nearest match. We then conclude that the time-series is expected to follow sequence 4 in the coming days. The sequence thus predicted would have practical value in the sense it helps users with the knowledge about the pathway to reach certain targets. This has important consequences in an economic time-series as it offers prediction to reach a definite target state (such as bullish state with high rise towards a saturated state) or a possible fall-off of the time-series to a deep down-valley (called bearish state), indicating a significant reduction in the price of a stock index item.

We also employ supervised classification in a time-series prediction problem. Typically, a time-series is partitioned vertically into equal sized interval, and each data point is categorized into one partition. Next we develop prediction rules by developing a transition from $P_i \rightarrow P_j$ for two consecutive data points at day r and day r + 1 in the time-series, where the time-series data on day r and r + 1 fall in partitions $P_i$ and $P_j$ respectively. Now, suppose we have a time-series containing

10,000 data taken on a daily basis. We obtain the transition rules for each successive pair of days and thus we have 9999 rules. We represent these rules by neural mapping, so that for a given time-series value at day r, we can obtain the data at day r + 1. The question that now arises as to how to store the training instance, so as to efficiently predict the next time-series data point. In this book, we propose a new approach to time-series prediction using an intelligent strategy to fire multiple rules: $P_i \rightarrow P_{j,}$ $P_i \rightarrow P_{k,}$ $P_i \rightarrow P_r$ having a common premise (representing a partition $P_i$) using multiple pre-trained neural networks with middle of the partitions $P_i$ and $P_{j,}$ $P_i$ and $P_{k,}$ $P_i$ and $P_r$ as the input-output component of training instances and take the average of the predicted outcomes obtained from these neural networks as the predicted time-series value.

## 1.5 Scope of Machine Learning in Time-Series Prediction

Neural nets play a vital role in learning the dynamic behavior of a time-series. The learning process in a neural network is accomplished by suitably selecting a set of cascaded non-linearity to represent the complex functional relations between the input and the output training instances supplied to the network. Neural network based learning is advantageous to Traditional Least Min Square (LMS) approach used in curve fitting [16] because of the pre-assumption of a fixed functional form of the latter rather than a variable functional form as used in the former. The variable functional architecture of the neural network offers the freedom to autonomously adapt its parameters in the required direction to appropriately tune the functional form of the network to produce the desired output for known inputs.

Several well-known approaches to supervised neural learning algorithms are prevalent in the literature. The most popular among these techniques are Back-propagation (BP) Learning algorithm [17], that adjusts the connectivity weights of neurons (single unit, comprising a weighted summer followed by a non-linearity) in the network layer-wise, starting from the last layer with an aim to minimize the error signals generated at the output layer for each input vector of the network. The error signal here is defined by taking the component-wise difference between the desired and the computed output vectors. The learning algorithm used in Back-propagation algorithm is designed following the well-known steepest descent learning policy that searches the direction of the shallowest gradient at a given point on the error surface of weights.

Among the other well-known algorithms used for supervised learning is Linear Support Vector Machine (LSVM) classifier that optimally segregates the input data space into two regions by straight line boundaries with sufficient spacing between the boundaries. Several variant of the basic LSVM are found in the literature [18]. The popular approaches used are kernelized SVM, where kernel functions are used to project the data points in new dimensions, thereby segregating the linearly inseparable data point by LSVM after kernel transformation. Besides BP and SVM there exist a lot many supervised neural classifiers, such as Radial Basis function Neural Net (RBF NN) [19], cascaded fuzzy-BP combinations and many others, the detailed listing is not given here for out of context. We next present a few well-known neural approaches to time-series prediction.

Frank et al. in [20] consider heuristics to select window size and sampling rate for efficient time-series prediction. One simple way to model time-series is to express the time-series x(t) as a non-linear function of x(t − 1), x(t − 2), …, x(t − n); i.e., x(t) = f(x(t − 1), x(t − 2), …, x(t − n)). Usually researches attempt to predict f by considering fixed non-linearity.

In Fig. 1.1, we consider a Sigmoid type of non-linear function. The learning process here requires to adapt weights $w_1, w_2, \ldots, w_n$ such that

$$\begin{aligned} z &= |f(g) - x(t)| \\ &= |f(\sum_{i=1}^{n} w_i \cdot x(t-i) - x(t))| \end{aligned} \tag{1.1}$$

is minimized. This is done by weight adaptation policy using the well-known gradient descent learning approach, presented below:

$$w_i \leftarrow w_i - \eta(\partial z/\partial w_i) \tag{1.2}$$

here, $f(g) = 1/(1 + e^{-g})$, is the Sigmoid function.

In [21], Faraway and Chatfield used a two layered architecture, where the first layer comprises logistic neurons, while the second layer involves a neuron with linear activation function. For logistic neurons, we use a Sigmoid type non-linearity over a weighted summer. For example,

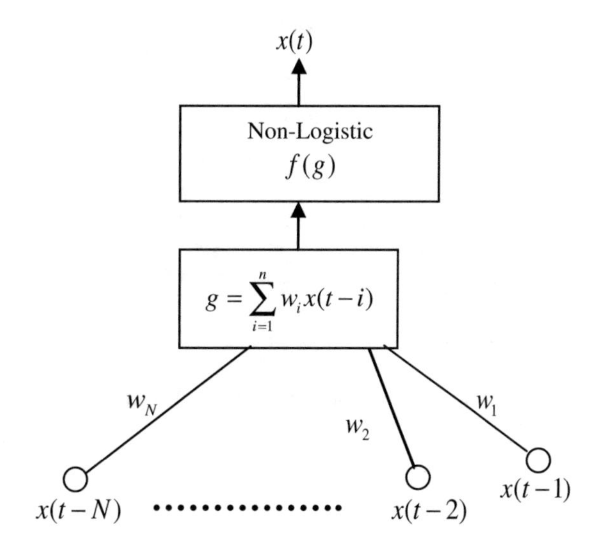


**Fig. 1.1** Time-series prediction using supervised neural net

$$y_t = f(\sum_{\forall i} x(t-i) \cdot w_j) \tag{1.3}$$

where, $f(z) = 1/(1 + e^{-z})$, is the Sigmoid function.

The output layer approximate $\hat{x}(t)$ is obtained by simply by a weighted summer followed by a nonlinear function. Mathematically,

$$\hat{x}(t) = h(\sum_{i=1}^{n} w_i \cdot y_i) \tag{1.4}$$

where, $h(x) = k \cdot x$, a linear function of x. In the present problem we used to predict $w_i$ and $w_j$,∀i, ∀j; so that $\hat{x}_t$ approximates $x(t)$ (Fig. 1.2).

Meesad et al. in [22] proposed a novel approach for stock market time-series prediction using support vector regression model. The model employs slack variables $\xi_i$ and $\xi_i^*$ for $i = 1$ to n, in order to measure the training samples outside the $\xi-$ sensitive zone [23, 24]. The regression model is given by

$$\phi(w, \xi) = (1/2) \cdot ||w|| + c \cdot \sum_{i=1}^{n} (\xi_i + \xi_i^*), \tag{1.5}$$

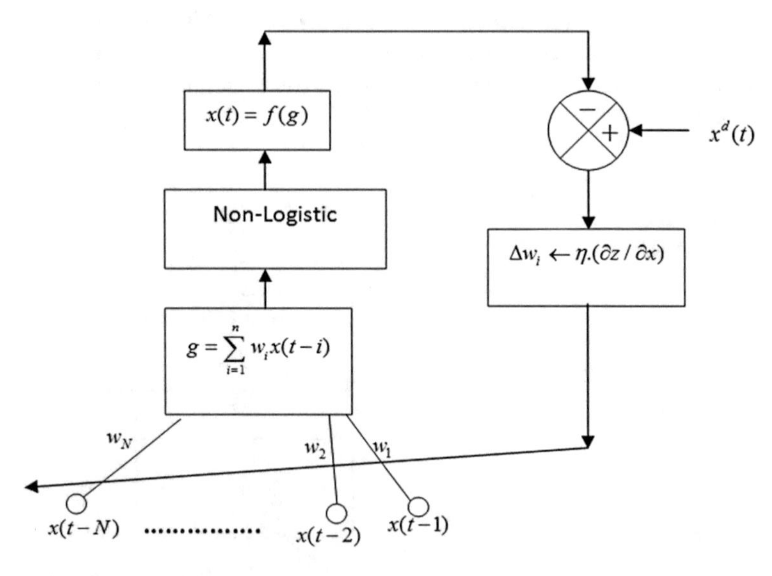


**Fig. 1.2** Weight adaptation using gradient descent learning

where the stable variables and $\xi_i^*$ fix the band of $||y_i - f(x_i - c)||$. Formally we have to fix $\xi_i^*$ by

$$Min(y_i - f(x_i, w)) \leq \xi_i + \xi_i^*$$
$$Min f(x_i, w) - y_i \leq \in + \xi_i^*$$

subject to

$$\xi_i \geq 0$$
$$\xi_i^* \geq 0$$

The above problem is transformed to its dual and solved using the Lagrange's multiplier method. The weights w thus obtained are used during prediction phase.

Li and Deng et al. in [25] proposed a novel approach to predict turning points in a chaotic financial time-series. According to them, in a given time-series $x_t \in R$, $t = 1, 2, 3, \ldots, n$, we have turning points $x_l$ (peak or trough) such that the change (increase/decrease) in the series at $t = l$ exceeds a specific percentage within $p$ steps (Fig. 1.3).

Based on the turning points obtained from the time series, an event characterization function, also called turning indicator $T_\gamma$ (t) is activated, where

$$\begin{aligned} T_\gamma(t) &= 1, \text{ if } x(t) \text{ is a peak,} \\ &= 0, \text{ if } x(t) \text{ is a trough.} \end{aligned}$$

The time series can be predicted at all the time points by linear interpolation.

Kattan et al. [26] introduced a novel technique for prediction of the position of any particular target event in a time series. Here, the authors determined a training set

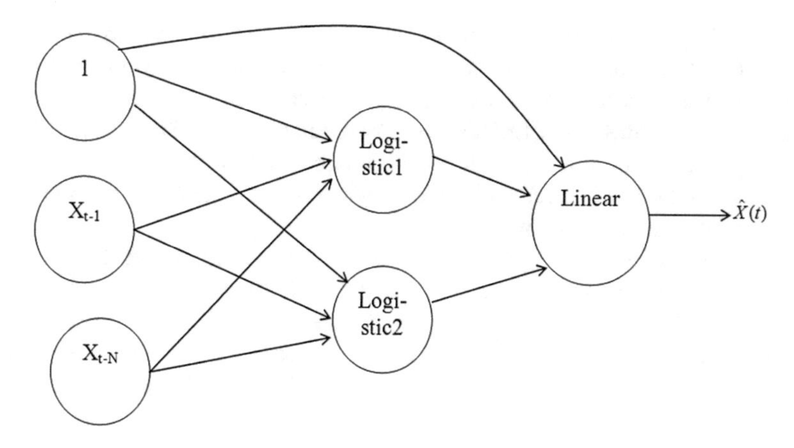


**Fig. 1.3** Feed-forward neural net work with look ahead

$$s = \{(V_a)|a = 0...n\}$$

where each $V_a$ is a time series vector such that $V_a = \{(x_t)|t = 0, 1, ..., t_{max}\}$. Here $t_{max}$ is the number of data points of the time series.

The work employs a *divide and conquer strategy* to classify frequency samples into different bins $\{(b_i), i = 0, 1, 2, ..., t_{max}\}$, where each bin is a subset of s. A genetic programming [27] is used to extract symmetrical features from each $b_i$'s members, and k-means clustering is employed to cluster $b_i$'s. After clustering is over, genetic programming is used as 'event detection' to identify the time slot in a linear time-series. The authors also compared the relative merit of their algorithm with existing algorithms including radial basis function neural nets, linear regression and polynomial regression.

The work by Miranian [28] employs a local neuro-fuzzy approach realized with least square support vector machines (LSSVM). The said LSSVM utilizes hierarchical binary tree (HBT) learning algorithm for time-efficient estimation of the local neuro-fuzzy model. Integration of LSSVMS into local neuro-fuzzy model network along with the HBT learning algorithm provides high performance approach for prediction of complex non-linear time-series.

Yan in [29] attempted to employ a generalized model of regression neural network (GRNN) with one design parameter and fast learning capability for time-series forecasting. The most attractive aspect of this research is to use multiple GRNN for forecasting of business time-series. The proposed model of GRNN is experimentally focused to be very robust with respect to design parameters, which makes it appropriate for large scale time-series prediction.

Lee et al. in [30] introduced Japanese candlestick theory as an empirical model of investment decision. It is preferred that the candlestick patterns represent the psychology of the investors and thus help the internet planners with a suitable advice on investment. Fuzzy linguistic labels are used to model the vagueness of imprecise candlesticks. The proposed fuzzy model transforms the candlesticks data into fuzzy patterns for recognition. The method developed is applied to financial time-series prediction.

In [31] Coyle et al. present a novel approach for feature extraction from EEG signals for left and right hand motor imageries. They employed two neural nets to perform one step ahead of prediction, where one neural net is trained on right motor imagery and the other on left motor imagery. Features are derived from the mean squared power of the prediction error. EEG signal is partitioned into equal sized intervals, and the EEG features of the next window are determined from the EEG features of the current window.

In Dash [32], the authors proposed a learning framework, where they used self adaptive differential harmony searched based optimized extreme learning machine (SADHS-OLEM) for single hidden layered feed-forward neural network. The learning paradigm introduced takes advantage of generalization ability of extreme learning machines with global learning potential for SADHS. The proposed technique has successfully been realized for close price and volatility prediction of five different stock indices.

There exist extensive literature on artificial neural network for time series prediction using feed-forward network, Boltzmann machines and deep belief network [33]. The work by Hrasko et al. [34] introduced a novel approach for time-series prediction using a combination of Gaussian Bernoli Restricted Boltzmann machine and the back-propagation algorithm.

## 1.6 Sources of Uncertainty in a Time-Series

A time-series apparently looks like a sequence of random valued data in a finite range, ordered over time. Because of this randomness, prediction of a time-series at the next time point from its current or preceding time-point values is difficult. Thus, for two equal time-series values: $x(t_1)$ and $x(t_2)$ occurring at time-point $t_1$ and $t_2$, it is not guaranteed that the next time point values of the series $x(t_1 + 1)$ and $x(t_2 + 1)$ need not be equal. To handle the non-deterministic behavior of the time-series, researchers usually do not take the risk to predict absolute next point value of the series, rather they offer a small range of the predicted value. This is realized by dividing the dynamic range of the time-series into a fixed number of partitions, usually of equal length, which reduces the uncertainty in the prediction as the prediction refers to identifying a partition containing the next time-point value instead of an absolute value.

A close inspection of a partitioned time-series now reveals the dependence relationships between the partitions containing the time-point value before and after a feasible time-point t present in the series. These relationships are generally referred to as prediction rules. It is indeed important to note that there exist uncertainty in the prediction rules: $P_i \rightarrow P_j$ and $P_i \rightarrow P_k$ where $P_i$, $P_j$ and $P_k$ are three partitions, where the transition $P_i$ to $P_j$ takes place around time-point $t_1$, while the transition $P_i$ to $P_k$ takes place around a second time-point $t_2$. Now, suppose, we like to predict the time-point value of the series at time $t' + 1$, where $t'$ falls in partition $P_i$. Now, in which partition should $x(t' + 1)$ lie? This is indeed difficult to say. Statisticians may favor the one with higher probability of occurrence. That is they would say $x(t' + 1)$ would lie in partition $P_j$ if $prob(P_j/P_i) > prob(P_k/P_i)$, where $prob(P_j/P_i)$ indicates the conditional probability of $P_j$ assuming the prior occurrence of $P_i$ around a given time-point $t'$. The above example illustrates that there exists uncertainty in prediction of a time-series for possible non-determinism of the extracted rules.

Another important aspect that influences a time-series prediction is secondary factor, which usually is hard to ascertain as its clarity is not visible in most circumstances. For example, a large fall-off in the close price of the DOW JONES time-series on day t may cause a significant fall-off in the TAIEX time-series on day t + 1. If this phenomenon is known, we call the DOW JONES time-series as the secondary factor for the main-factored time-series TAIEX. In absence of any knowledge of the secondary factors, the prediction of a main-factored time-series remains uncertain.

The third important issue responsible for uncertain prediction of a time-series is its frequency of occurrence in a partition. It is observed that a time-series remains in fewer partitions most frequently than the rest. Thus its prediction probability is higher in fewer partitions. However, because of non-stationary characteristics of the series, the partitions having higher frequency of occurrence vary over time, and thus extraction of the highly probable partitions at a given time is uncertain. There are other sources of uncertainty in a time-series. However, in the present book we consider these three types of uncertainty in time-series modeling and forecasting.

## 1.7 Scope of Uncertainty Management by Fuzzy Sets

Fuzzy sets are widely being used for uncertainty management in expert systems. Because of multiple sources of uncertainty in the prediction of a time-series, the logic of fuzzy sets can be used to handle the problem. Fuzzy logic primarily is an extension of classical logic of propositions and predicates. In propositional/ predicate logic, we use binary truth functionality to represent the truth value of a proposition/predicate. Because of strict binary truth functionality, propositional/ predicate logic fails to express the uncertainty of the real-world problems. In fuzzy logic, the truth values of a fuzzy (partially true) proposition lies in the closed interval of [0, 1], where the binary digits: 0 and 1 indicate the completely false and totally true.

Consider, for instance, a fuzzy production rule: *if x is A then y is B*, where x and y are linguistic variables and A and B are fuzzy sets in respective universe U and V respectively. The connectivity between x is A and y is B is represented by a fuzzy implication relation,

$$
R(x, y) = \begin{array}{c|cccccc}
{}_{x}\diagdown^{y} & y_1 & y_2 & \cdots & y_j & \cdots & y_m \\
\hline
x_1 & \cdots & \cdots & & \cdots & & \cdots \\
x_2 & \cdots & \cdots & & \cdots & & \cdots \\
\cdots & \cdots & \cdots & & \cdots & & \cdots \\
x_i & \cdots & \cdots & & R(x_i, y_j) & & \cdots \\
 & \cdots & \cdots & & \cdots & & \cdots \\
x_n & \cdots & \cdots & & \cdots & & \cdots
\end{array}
$$

where $x \in \{x_1, x_2, \ldots\ldots, x_n\} = X$, $y \in \{y_1, y_2, \ldots\ldots, y_m\} = Y$ and R(x, y) denotes the strength of fuzzy relation for $x = x_i$ and $y = y_j$, satisfying the implication between x is A and y is B.

Different implication functions are used in the direct use to describe the fuzzy if-then connectivity. A few of the well-known implication relations are Mamdani, Lucksiwcz and Diens-Rescher [35, 36] relations.

Mamdani implication relation

$$\begin{aligned} \mathrm{R}(x, y) &= \mathrm{Min}(\mu_A(x_i), \mu_B(y_j)) \\ \text{or } \mathrm{R}(x, y) &= \mu_A(x_i) \times \mu_B(y_j) \end{aligned} \tag{1.6}$$

Lukasiewicz implication relation

$$\mathrm{R(x,\ y)} = \mathrm{Min}\ [1, (1 - \mu_A(x_i) + \mu_B(y_j))] \tag{1.7}$$

Deins-Rescher implication relation

$$\begin{aligned} \mathrm{R(x, y)} &= \mu_{\bar{A}}(x_i) \vee \mu_B(y_j) \\ &= \mathrm{Max}\ [\mu_{\bar{A}}(x_i), \mu_B(y_j)] \end{aligned} \tag{1.8}$$

where $\bar{A}$ denotes the complementation of the fuzzy set A. In fuzzy reasoning, we typically have a fuzzy relation R(x, y) for a given fuzzy production rule: if $x$ is $A$ then $y$ is $B$ and a fuzzy proposition $x$ is $A'$, we need to infer the membership function of $y$ is $B'$.

$$\begin{aligned} \mu_{B'}(y) &= \mu_{A'}(x) \, o \, R(x, y) \\ &= \underset{x}{\mathrm{Max}}[\mathrm{Min}(\mu_{A'}(x), R(x, y))] \\ &= \underset{x}{\mathrm{Max}}[\mathrm{Min}(\mu_{A'}(x_i), \mu_R(x_i, y))], \quad \textit{for } x_i \in X, \ \forall i = 1 \text{ to } n, \\ &= \underset{x}{\mathrm{Max}}[\mathrm{Min}\{\mathrm{Min}(\mu_{A'}(x_i), \mu_A(x)), \mu_{B'}(y)\}], \quad \forall x \end{aligned} \tag{1.9}$$

The principle of fuzzy reasoning is illustrated below graphically using Mamdani implication relation.

Occasionally, a fuzzy rule includes two or more propositions in the antecedent. For instance consider the fuzzy production rule: if x is A and y is B then z is C, where A, B, C are fuzzy sets and x, y, z are the fuzzy linguistic variables. Consider fuzzy observations x is A′ and y is B′.

To get back to the real world we need to defuzzify the inference $\mu_{B'}(y)$ in Figs. 1.4 and 1.5 by the following procedure

$$\bar{y} = \frac{\sum_{\forall y} \mu_B(y) \cdot y}{\sum_{\forall y} \mu_B(y)} \tag{1.10}$$

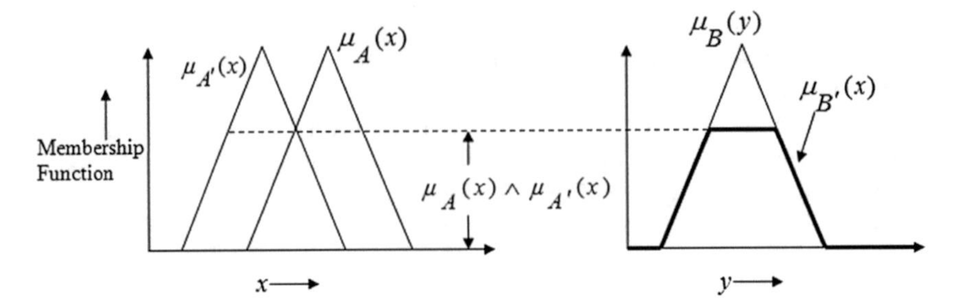


**Fig. 1.4** Inference generation mechanisms

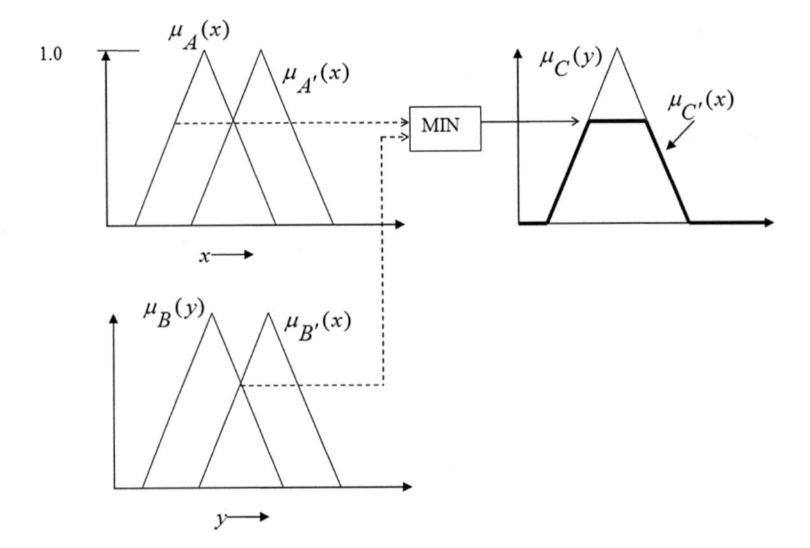


**Fig. 1.5** Inference generation using type-1 fuzzy reasoning

and

$$\bar{z} = \frac{\sum_{\forall z} \mu_C(z) \cdot z}{\sum_{\forall z} \mu_C(z)}. \tag{1.11}$$

Fuzzy reasoning involves three main steps: (i) fuzzification, (ii) inference generation and (iii) defuzzification. In the fuzzification step, we represent a fuzzy linguistic variable $x$ into membership of $x$ to belong to a fuzzy set A. In other word, fuzzification introduces a mapping: $x$ to $\mu_A(x)$. This is done by a specially designed membership function $\mu_A(x)$. Fuzzy reasoning and defuzzification steps have been briefly outlined above.

## 1.8 Fuzzy Time-Series

Prediction of a time-series from its current and/or past samples is an open area of current research. Unfortunately, traditional statistical techniques or regression-based models are not appropriate for prediction of economic time-series as the external parameters influencing the series in most circumstances are not clearly known. Naturally, prediction of a time-series requires reasoning under uncertainty, which can be easily performed using the logic of fuzzy sets. Researchers are taking keen interest to represent a time-series using fuzzy sets for an efficient reasoning. One well-known approach, in this regard, is to partition the dynamic range of the time-series into intervals of equal size, where each interval is described by a fuzzy set. Several approaches of representation of the partitions of a time-series by fuzzy sets are available in the literature [37–95]. One simple approach is to represent each partition by a membership function (MF), typically a Gaussian curve, with its mean equal to the centre of the partition and variance equal to half of the height of the partition. Thus each partition $P_i$ can be represented as a fuzzy set $A_i$, where $A_i$ is a Gaussian MF of fixed mean and variance as discussed above.

Choice of the membership function also has a great role to play on the performance of prediction. For example, instead of Gaussian MF, sometimes researchers employ triangular, left shoulder and right shoulder MFs to model the partitions of a time-series. Selection of a suitable MF describing a partition and its parameters remained an open problem until this date.

**Defining a Fuzzy Time-series**

Let $Y(t) \subset R$, the set of real numbers for $t = 0, 1, 2, \ldots$, be a time-varying universe of discourse of the fuzzy sets $A_1(t)$, $A_2(t)$, ... at time t, where $A_i(t) = \{y_i(t)/\mu_i(y_i(t))\}$, for all i and $y_i(t) \in Y(t)$ We following [37, 38] define a fuzzy time-series $F(t) = \{A_1(t), A_2(t), \ldots\}$.

The following observations are apparent from the definition of fuzzy time-series.

1. In general, a fuzzy time-series F(t) includes an infinite number of time-varying fuzzy sets $A_1(t)$, $A_2(t)$,..
2. The universe of discourse Y(t) is also time-varying.
3. F(t) is a linguistic variable and $A_i(t)$ is its possible linguistic values (fuzzy sets), for all i.

For most of the practical purposes, we re-structure the above definition of fuzzy time-series by imposing additional restrictions on its parameters. First, we consider a finite number of time-invariant fuzzy sets $A_1$, $A_2$, ..., $A_n$. Second, the universe of discourse Y is time-invariant.

*Example 1.1* Consider a close price time-series for the TAIEX data for the period 2000–2001, which is horizontally partitioned into 7 intervals: $U_1$, $U_2$, ..., $U_7$, where the union of these partitions represent the dynamic range, called the *Universe of discourse* U of the time-series. In other words,

$$U = \cup_{i=1}^{7} U_i. \tag{1.12}$$

Let the fuzzy sets used to represent the time-series be EXTREMELY-LOW, VERY-LOW, LOW, MODERATE, HIGH, VERY-HIGH and EXTREMELY-HIGH, which are hereafter described by $A_1$, $A_2$, …, $A_7$ respectively. Typically, we aim at representing the 7 partitions by 7 fuzzy sets with overlap in the data points. The overlapping appears due to inherent fuzziness of the fuzzy sets $A_1$, $A_2$, …, $A_7$ which have a common universe U and thus any member $y_i(t)$ of U falls in all the fuzzy sets with a certain membership lying in [0, 1]. Figure 1.6 provides the partitions and the respective fuzzy sets of the TAIEX time-series. In Table 1.1, we consider a set of 10 points on the series and the assignment of fuzzy sets to those points.

The definition of fuzzy time series introduced by Song and Chissom [37–40] has widely been used by subsequent researchers with suitable membership functions of their choice. In [41–50] Chen et al., the authors considered three valued membership functions of any point on the time-series to lie in a given fuzzy set $A_i$. For

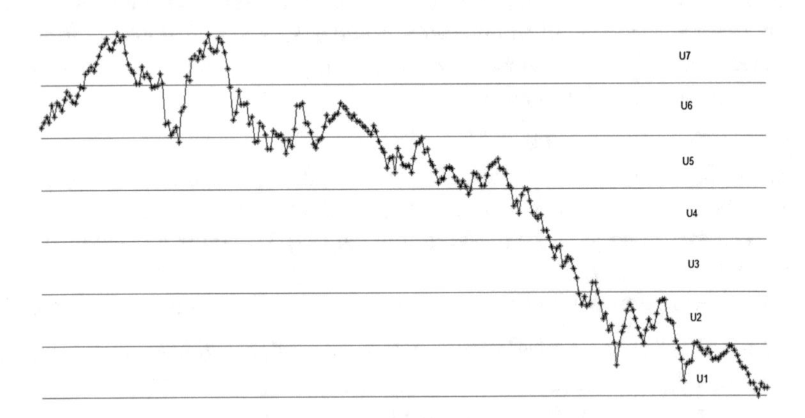


**Fig. 1.6** The TAIEX close price divided into 7 *horizontal partitions*, each describing a fuzzy set

**Table 1.1** Membership assignment to data points lying in partitions $U_1$ through $U_7$

| Fuzzy Sets | Partitions | | | | | | |
|---|---|---|---|---|---|---|---|
| | $U_1$ | $U_2$ | $U_3$ | $U_4$ | $U_5$ | $U_6$ | $U_7$ |
| $A_1$ | 1 | 0.5 | 0 | 0 | 0 | 0 | 0 |
| $A_2$ | 0.5 | 1 | 0.5 | 0 | 0 | 0 | 0 |
| $A_3$ | 0 | 0.5 | 1 | 0.5 | 0 | 0 | 0 |
| $A_4$ | 0 | 0 | 0.5 | 1 | 0.5 | 0 | 0 |
| $A_5$ | 0 | 0 | 0 | 0.5 | 1 | 0.5 | 0 |
| $A_6$ | 0 | 0 | 0 | 0 | 0.5 | 1 | 0.5 |
| $A_7$ | 0 | 0 | 0 | 0 | 0 | 0.5 | 1 |

instance, if a point lies in a partition i, we assign a full membership of one to the point to lie in fuzzy set $A_i$. If the point belongs to the next neighbors of partition i, we assign it a membership of 0.5 to lie in fuzzy set $A_i$. Further, if a point lies in any other partition, we assign it a membership of zero to belong to partition i.

Choice of membership functions is an important issue in fuzzy time-series prediction. In one of the recent papers [96], the authors consider left shoulder, right shoulder and triangular membership functions to describe the fuzziness involved in a time-series. In a recent work [62], Bhattacharya et al. modeled the fuzziness in the partitions using Gaussian membership functions. They considered all possible contiguous data points in the time-series within a partition by a Gaussian membership function, and later combined all such Gaussian membership functions for each contiguous block of data points in the partition using an interval type-2 fuzzy set [97].

### *1.8.1 Partitioning of Fuzzy Time-Series*

Partitioning is one of the fundamental steps in time-series prediction. The essence of partitioning arises due to our inability to accurately predict the time-series from its current and previous sample values. Partitioning helps to divide the dynamic range (signal swing) of the time-series into small intervals. This helps in prediction as we can predict the possible partition containing the next time-point rather than the next time-point itself.

Early research on time-series prediction considered uniform partitioning, where the interval widths were set equal. Uniform partitioning is acceptable to many researchers for its simplicity. It works with fewer parameters: partition width, beginning of the first partition (containing the lowest time-series value) and the end of the last partition. Usually the beginning of the first partition is set slightly below the global minimum of the time series within a finite time bounds, whereas the end of the last partition is set slightly above the global maximum. The extent of bounding partitions below and above the dynamic range of the time-series is selected intuitively. The extensions on either side of the dynamic range are required to allow possible occurrences of the predicted partition outside the dynamic range. The extended dynamic range is generally referred to as the universe of discourse.

Selection of partition width and partition-count are important issues. In [55], authors attempted to capture at least half of the consecutive time-point values in distinct partitions. Selection of a suitable partition-width capable of satisfying the above criterion is an optimization problem. However, the early researchers attempted to solve it in a heuristic approach. Huarng suggested two alternative proposals to handle the problem, In the first proposal, he considered the greatest value smaller than the median of the absolute first differences as the width of the partition. His second proposal is concerned with defining the average of absolute first differences of the time-series data points and fixing half of the average as the partition-width. He has experimentally demonstrated that the latter proposal comes up with better prediction results with respect to the former.

Partitioning a time-series into equi-spaced intervals is not free from flaws. Some of the disadvantages of uniform partitioning include (i) one or more partitions may be vacant containing no data points, and thus may not carry any significance for partitioning, (ii) the data density in the partitions may be non-uniform, which may give rise to erroneous prediction of the partition on the next day from today's time-series value/partition number, (iii) one or fewer partitions may include distributed fragments of time-series data, while other may include only continuum data for small intervals of time. The distributed fragments of time-series data carry much information of the time-series, and the partition containing such data describes certain characteristics of the time-series, while partitions containing small continuum data only include local characteristics of the time-series.

Although this is the simplest approach to define partitions in a time-series, the said method is not free from flaws. A few common shortcomings of uniform partitioning include: (i) one or more partitions may be vacant containing no data points, and thus may not carry any significance for partitioning, (ii) the data density in the partitions may be non-uniform, which may give rise to erroneous prediction of the partition on the next day from today's time-series value/partition number, (iii) one or fewer partitions may include distributed fragments of time-series data, while other may include only continuum data for small intervals of time. The distributed fragments of time-series data carry much information of the time-series, and the partition containing such data describes certain characteristics of the time-series, while partitions containing small continuum data only include local characteristics of the time-series.

Huarng proposed two simple but interesting algorithms [61] for uniform partitioning of a time-series. In both the algorithms, the primary motivation is to set the partition-width so as to accommodate the local fluctuation (obtained by taking the absolute difference between two successive time-series data points) of the time-series within a partition. The first algorithm assigns the highest value immediately lower than the median of the absolute differences as the partition width. The second algorithm, on the other hand, evaluates the average of the local fluctuations over a given time-frame, and takes half of the average value as the partition-width.

Extensive research on time-series partitioning has been undertaken over the last two decades, which could overcome a few of the above limitations of uniform partitioning. The principle undertaken to overcome the limitations includes intelligent non-uniform partitioning of the time-series, so as to satisfy certain criteria independently or jointly. One of the popular but simple density based clustering is due to Singh and Borah [98], where the authors attempted a two-step procedure to partition the time-series. In the first step, they partition the dynamic range of the time-series into two blocks of equal width. In the second step, they uniformly partition each block based on their data density. In other words, the block with higher data density is partitioned into more number of equal sized intervals than the block with lower data density.

### 1.8.2 *Fuzzification of a Time-Series*

Predicting a time-series using fuzzy logic requires three primary steps: fuzzification (also called fuzzy encoding [99]), fuzzy reasoning and defuzzification (fuzzy decoding [51], [100]). The fuzzification step transforms a time-series into fuzzy quantity, described by a membership value (of the sample points of a time-series) along with the sample value in one or more fuzzy sets. For instance, let $A_1$ (VERY SMALL), $A_2$ (SMALL), $A_3$ (MEDIUM), $A_4$ (LARGE) and $A_5$ (VERY LARGE) be linguistic values/fuzzy sets in a given universe, defined in the dynamic range of the time-series. Usually, the fuzzy sets are represented by membership curves lying in [0, 1] with respect to time-series sample values. Fuzzifier thus refers to construction of the membership curves, so that an unknown time series value can be directly transformed to a membership value by using a membership curve.

The apparent question is how to construct a membership function. This, in fact, is designed intuitively, using the background knowledge about the logical meaning of the linguistic values. For instance, the adjective LARGE will take up larger memberships for larger values of the time-series samples, and so can be best described by an exponential/linear rise to a certain value of the linguistic variable followed by a flat level. Any standard mathematical form of the membership function that satisfies the above criterion qualifies as the membership function for the given linguistic value. Most commonly, the VERY LARGE/LARGE is represented by a Left shoulder, an exponential rise with a flat end, special type parabola, and the like [101]. The MEDIUM linguistic value is described usually by an isosceles triangle, trapezoid or Gaussian type membership function, while the SMALL linguistic value is described by a right shoulder or an exponential decay. Figure 1.7 provides the structures of the above mentioned MFs.

Chen et al. in [41–50, 52, 53], considered a simplified approach to describe a MF by ternary membership values: 0, 0.5 and 1. Such representation has the primary advantage of reducing subsequent computations in reasoning and inference generation stage. One typical such MF representing CLOSE_TO_THE_CENTRE_OF_PARTITION $P_i$ is indicated in Fig. 1.7.

In case the time-series value lies in partition $P_i$ we assign a membership of 1.0 to that data point. If it falls in one of the two closest partitions on either side of $P_i$, we assign the data point a membership equal to 0.5. If the data point falls in any other partition except in partition $P_i$ and its immediate neighbors, we assign the data point a membership equal to 0. In most of the subsequent works, researchers follow the above principle of membership function assignment in a time-series.

In [62] Bhattacharya et al., the authors attempted to describe the time-series data points in a partition by a Gaussian MF. Let the partition $P_i$ includes the data points $c(t_p)$, $c(t_q)$, …, $c(t_r)$. Let the mean and variance of these data points be $m_i$ and $s_i^2$ respectively. We then describe the data points in partition $P_i$ by a Gaussian curve N $(m_i, s_i)$ given below.

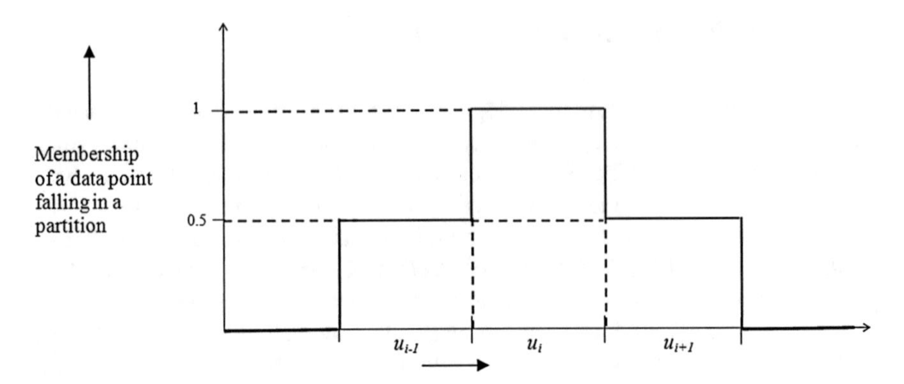


**Fig. 1.7** One way of defining a fuzzy time-series

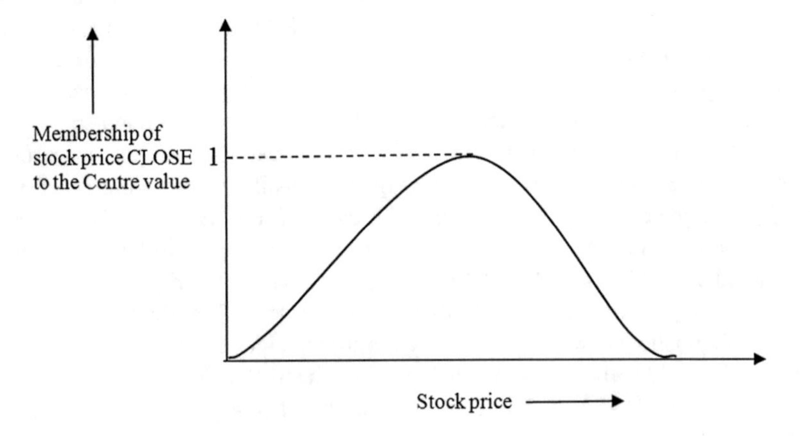


**Fig. 1.8** The representation of partition $P_i$ by a Gaussian membership function with its centre value as the centre of the Gaussian curve

$$N(m_i, s_i) = (1/s_i\sqrt{(2\pi)}) \cdot \exp[(-(c(t) - m_i)^2]/2 \cdot s_i^2 \tag{1.13}$$

An alternative way to represent the time-series by a Gaussian membership function is to fix the peak of the Gaussian at the centre of partition $P_i$, allowing its excursions to both the left and right extremities of the next partition. Figure 1.8 provides a schematic representation of such membership function. Alternatively, this can be represented by a triangular MF (Fig. 1.9) with the peak at the centre of the i-th partition and fall-off of the membership around the peak covering the neighborhood partitions.

Assigning a linguistic value to each data point of the time-series is essentially a two-step process:

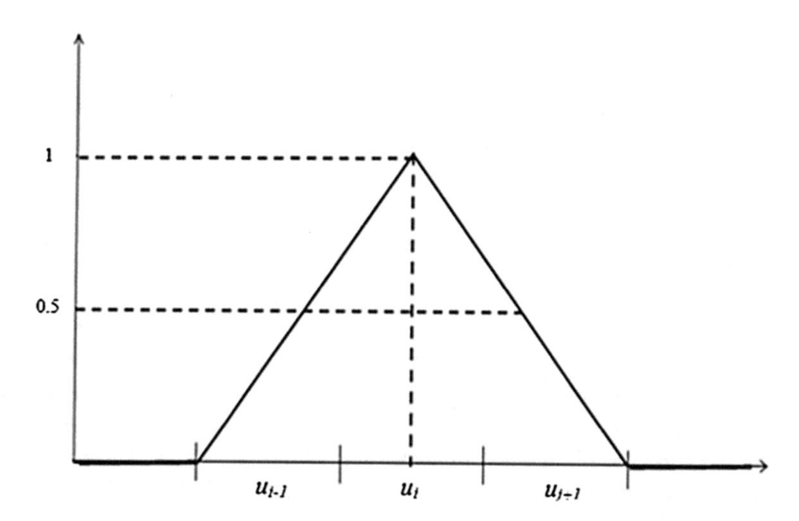


**Fig. 1.9** The representation of partition $P_i$ by a triangular membership function

1. According to the membership functions defined for the fuzzy sets, compute the membership value of the data point for each of the fuzzy sets defined on the universe of discourse.
2. Assign the fuzzy set for which the data point has maximum membership value, as the linguistic value for the data point.

In [102], the authors attempted to determine membership functions to describe a time-series by employing fuzzy c-means clustering algorithm. The motivation in this paper is to represent a data point $\hat{d}_k$ in the time-series by m one dimensional clusters with an associated membership in each cluster, such that the following two conditions jointly hold:

$$\sum \forall \mathrm{i}\ \mu_{\mathrm{Ci}}(\hat{d}_k) = 1 \tag{1.14}$$

and $0 < \mu_{\mathrm{Ci}}(\hat{d}_k) < 1$ for at least one data point $\hat{d}_k$ in a given cluster Ci.

The first criterion indicates that the membership of a data point $d_k$ in all the clusters (partitions) together is equal to one. This, in other words, implicates that if the data point belongs to a specific cluster with high membership close enough to one, its membership in all other clusters must be very small, close to zero.

The second criterion indicates that for some k, the data point $\hat{d}_k$ should have a non-zero membership to belong to a given cluster Ci, failing which the cluster Ci would have no existence, which is contrary to the assumption of m clusters. This explains the part of the second condition: $0 < \mu_{\mathrm{Ci}}(\hat{d}_k)$. On the other hand, the second inequality: $\mu_{\mathrm{Ci}}(\hat{d}_k) < 1$ indicates that $\hat{d}_k$ has partial membership in other clusters too, failing which (i.e., $\mu_{\mathrm{Ci}}(\hat{d}_k) = 1$) the fuzziness of $\hat{d}_k$ to lie in cluster Ci is lost.

After fuzzification by one dimensional FCM clustering, the authors obtained the highest degree of membership of each data point of the time-series in cluster Ci for all i = 1 to m. The bounding partitions of cluster Ci are formed by enclosing those supporting data points with highest membership in that cluster.

## 1.9 Time-Series Prediction Using Fuzzy Reasoning

One primary motivation of time-series prediction by fuzzy logic based reasoning is to determine the implication functions that exist between the membership function of a time point and its next time point in a time series. For instance, consider a close price time-series c(t) where $\mu_A(c(t-1))$ and $\mu_B(c(t))$ denote the respective membership values of the close price in fuzzy sets A and B. We like to express the implication relation from "c(t − 1) is A" to "c(t) is B" by a fuzzy relation R(c(t − 1), c(t)). Now, suppose we have m partitions, and suppose the data points in a partition is represented by a suitable fuzzy set. Thus for m partitions P1, P2, …, Pm, we have m fuzzy sets A1, A2, …, Am. To extract the list of all possible implications, we thus require to mark the transitions of the time-series at all time points t − 1 and t and then identify the appropriate fuzzy sets to which c(t − 1) and c(t) belong with the highest memberships.

Let c(t − 1) and c(t) have highest memberships in $A_p$ and $A_q$ for certain value of $t = t_i$. Then we would construct a fuzzy relation for the rule: If c(t − 1) is $A_p$ Then c (t) is $A_q$. The question then naturally arises as to how to construct the implication relation. In fact there exists quite a few implication relations, a few of these that are worth mentioning includes Mamdani (Min), Lukasiewicz and Diens-Rescher implications. Most of the existing literature on fuzzy time-series, however, utilizes Mamdani (Min-type) implication for its simplicity, low computational complexity and publicity.

Once a fuzzy implication relation R(c(t − 1), c(t)) is developed, we would be able to predict the fuzzy membership value $\mu_B'(c(t))$ for the next time-series data point c(t) from the measured membership value $\mu_A'(c(t-1))$ of the close price at time c(t − 1). The dash over A′ and B′ here represents that the fuzzy set A′ and B′ are linguistically close enough to the respective fuzzy sets A and B respectively. The fuzzy compositional rule of inference has been used to determine $\mu_B'(c(t))$ by the following step:

$$\mu'_B(c(t)) = \mu'_A(c(t-1)) \text{ o } R \tag{1.15}$$

where o is a max-min compositional operation, which is similar to matrix multiplication operation with summation replaced by maximum and product replaced by minimum operators.

In the first successful work by Song and Chissom [37], the authors proposed a novel approach to determine R from a given university enrollment time-series. Given a n-point time-series, they evaluated R(c(t − 1), c(t)) for t = 1, t = 2, …, t = n − 1. Let the implication relations in order be $R_1, R_2, \ldots, R_{n-1}$. They define the time-invariant fuzzy implication relation

$$R = \cup_{i=1}^{n-1} R_i \tag{1.16}$$

The basic philosophy behind the above formulation of R lies in the assumption that the implication relation of the entire time-series is time-invariant. In later models, proposed by Song and Chissom [37–40], the authors considered a time-variant model of the fuzzy implication relations, where they considered a time-window of width w, over which the implication relation is constant, but it varies across the windows.

The primary motivation behind the study of any time-series is to develop a model with the potential for accurate prediction of the time-series. In case of a fuzzy time-series where each observation (or data point) is a fuzzy linguistic value, the model that is developed is primarily a fuzzy relationship between the current linguistic value (or fuzzy set) and those obtained in the past. In this regard, Song and Chissom [37–40] defined a first-order fuzzy relationship between two consecutive fuzzy time-series values, $F(t)$ and $F(t-1)$ as given by the following definition.

**Definition 1.1** If the fuzzy time-series observation at the *t*th instant, i.e., $F(t)$ is solely dependent on the fuzzy time-series observation at the $(t-1)$th instant $(F(t-1))$, then this dependency can be expressed in the form of a **first-order fuzzy relationship** $R(t-1, t)$ as given below

$$F(t) = F(t-1) \circ R(t-1, t). \tag{1.17}$$

It should be noted, that the operator "$\circ$" given in (1.17) is a generic fuzzy operator which can be defined in several ways. Song and Chissom [38], in their study chose the max-min composition fuzzy operator for developing the first-order fuzzy relationship $R(t-1, t)$. Based on the nature of the fuzzy relationship $R(t-1, t)$, a fuzzy time-series can be broadly classified into two major categories: (i) Time-variant fuzzy time-series, (ii) Time-invariant fuzzy time-series.

**Definition 1.2** If, the first-order fuzzy relationship $R(t-1, t)$ is independent of time $t$, i.e., for every time instant $t$, $R(t-1, t)$ holds a constant value (say $R$), then the fuzzy time-series concerned, is said to be **time-invariant**. On the contrary, if for any time-instant $t, R(t-1, t) \neq R(t-2, t-1)$, then the fuzzy relationship is time-dependent and the fuzzy time-series is called a **time-variant fuzzy time-series**.

Song and Chissom in their works [38–40] used both time-invariant and time-variant models to predict the student enrollments of the University of Alabama. For the time-invariant model, a single fuzzy relationship matrix $R$, is computed using all the past observations ($F(1)$–$F(t-1)$ for prediction of $F(t)$) of the fuzzy time-series, whereas, for the time-variant model, the general approach is to consider a fixed window size $w$, and develop the fuzzy relationship matrix $R(t-1, t)$, from the past $w$ observations only (i.e.,$F(t-w)$ to $F(t-1)$ for prediction of $F(t)$).

The first-order model is based on the assumption that the fuzzy time-series observation at the $t$th instant ($F(t)$) is caused solely by the observation at the $(t - 1)$ th instant. However, this assumption need not always be true and often it is seen that a time-series data point has dependencies on multiple past data points. To incorporate this concept in the case of a fuzzy time-series, a higher order model for fuzzy time-series was developed, where the dependency of $F(t)$ on a sequence of past fuzzy time-series observations is captured by the higher order fuzzy relationship matrix. For example, the fuzzy relationship matrix of an $m$th order fuzzy time-series is computed as follows.

$$F(t) = (F(t-1) \times F(t-2) \times \ldots \times F(t-m)) \circ R_m(t-m,\ t) \tag{1.18}$$

In (1.18), $R_m(t-m,\ t)$ denotes the $m$th order fuzzy relationship matrix for the fuzzy time-series $F(t)$. Apart from the model proposed by Song and Chissom [37–40], there have been a number of variants for modeling a fuzzy time-series with the motive of accurate prediction. For instance, Sullivan and Woodall [64] proposed both a time-invariant and a time-variant fuzzy time-series prediction methodology based on Markov models and applied their model for prediction of student enrollments at the University of Alabama. Given a vector of state probabilities $P_t$ at time-instant $t$, the same for time-instant $(t+1)$, $P_{t+1}$ can be computed as given below.

$$P_{t+1} = P_t \times R_m \tag{1.19}$$

$$P_{t+1} = P_t \times R_m^k \ (k = 1, 2, \ldots) \tag{1.20}$$

In (1.10) and (1.20) $R_m$ and $R_m^k$ are state transition matrices for the time-invariant and time-variant cases respectively and the '$\times$' operator is a matrix multiplication operator (not a fuzzy operator).

There exist certain disadvantages of computing a relationship matrix for prediction of a fuzzy time-series. One of the most conspicuous drawbacks is the computational load involved in calculating the fuzzy relationship matrix using fuzzy logical operators like the max-min composition operator. To solve this problem, Chen [41–50] proposed a simple first-order model for a fuzzy time-series using arithmetic operators. In the proposed model, firstly, each data point in the time-series is fuzzified to a particular linguistic value (i.e., the linguistic value associated with the fuzzy set for which the data point has maximum membership). Next, first-order fuzzy logical relationships (FLR) are computed from the fuzzified time-series where, for any time-instant $t$, if $A_i$ is the corresponding fuzzified value and $A_j$ is the fuzzified value for the $(t - 1)$th instant, then the first-order FLR obtained is

$$A_j \rightarrow A_i. \tag{1.21}$$

The third step includes grouping the above fuzzy logical relationships into fuzzy logical relationship groups (FLRG) based on common antecedents. For instance, the FLRs $A_j \to A_{i1}$, $A_j \to A_{i2}$ and $A_j \to A_{i3}$ can be grouped into the FLRG, $A_j \to A_{i1}, A_{i2}, A_{i3}$. It should be noted that in this model, multiple occurrences of a particular FLR have not been prioritized. In other words, even if a particular FLR occurs multiple times, it will only be considered once in its corresponding FLRG. The FLRGs are later defuzzified for prediction purposes. A detailed description of defuzzification techniques is given in a later section [64–95].

A simple drawback of the above model is that it assigns the same weight to each FLR irrespective of either the number of occurrences of the FLR or its corresponding chronological order. Yu [55–60] observed that the recurrence in the occurrence of FLRs should be exploited and proposed a model to consider weighted FLRs based on their chronological order. Let the sequence of FLRs in chronological order be as follows

$$\begin{aligned} FLR\ 1 &: A_i \to A_{j1} \\ FLR\ 2 &: A_i \to A_{j2} \\ FLR\ 3 &: A_i \to A_{j3}. \end{aligned}$$

The idea is to assign a higher weight to more recent FLRs as compared to older FLRs. Thus, FLR 1 is assigned a weight $w_1$ of 1, FLR 2 is assigned a weight $w_2$ of 2 and FLR 3 is assigned a weight $w_3$ of 3. In the next step, the weights are normalized, as

$$w_k = \frac{w_k}{\sum_{j=1}^{3} w_j} \tag{1.22}$$

so that each $w_k \in [0, 1]$. Hence, the FLRG obtained is $A_i \to A_{j1}, A_{j2}, A_{j3}$ with the corresponding weights assigned to each FLR. It should be noted that the consequents of the FLRG obtained using the above process need not be distinct. Since each consequent is weighted based on the sequence or chronology in which it arrived, it is considered separately. This feature is different from Chen's earlier model where multiple occurrences of the same consequent in FLRs was reduced to a single consequent in the FLRG. Another approach in the direction of weighted fuzzy relationships was proposed by Uslu et al. [54] where the FLRs are weighted based on their number of occurrences instead of their chronological order. An FLR having higher number of occurrences possesses a greater weight than an FLR with relatively lower number of occurrences [37–51].

There have been numerous approaches made by researchers to improve the existing models of fuzzy time-series. A very basic but important observation in this regard is that a time-series may be dependent on more than one factors. The general approach to developing prediction models for any time-series (fuzzy/conventional) is to predict the future values of the time-series based solely on knowledge obtained from the time-series values of the past. However, it is generally seen that better prediction results can be obtained if multiple factors on which a time-series is dependent are incorporated in the model. In this direction, Huarng [103] proposed a heuristic fuzzy time-series model where, in the first step, fuzzy relationships are obtained based on Chen's model [41–50]. In the second step, the obtained FLRs are refined based on heuristic knowledge obtained from secondary time-series. Let the initial FLRG obtained for the antecedent $A_i$ be $A_i \rightarrow A_{p1}, A_{p2}, \ldots, A_{pn}$. The FLRG is then further refined based on heuristic variables $x_1, x_2, \ldots x_k$ as follows.

## 1.10 Single and Multi-Factored Time-Series Prediction

Traditional techniques of time-series prediction solely rely on the time-series itself for the prediction. However, recent studies [96] reveal that the predicted value of the time-series cannot be accurately determined by the previous values the series only. Rather, a reference and more influential time-series, such as NASDEQ and DOWJONES can be used along with the original time-series for the latter's prediction.

Several models of forecasting have been developed in the past to improve forecasting accuracy and also to reduce computational overhead. There exist issues of handling uncertainty in business forecasting by partitioning the intervals of non-uniform length. The work undertaken by Li and Cheng [104] proposes a novel deterministic fuzzy time-series to determine suitable interval lengths. Experiments have been performed to test the feasibility of the forecasting on enrolment data in Alabama Universities. Experimental results further indicate that the first order time series used here is highly reliable for prediction and thus is appropriate for the present application.

Existing models of fuzzy time-series rely on a first order partitioning of the dynamic range of the series. Such partitioning helps in assigning fuzzy sets to individual partitions. First order partitions being uniform cannot ensure a large cardinality of data points. In fact, occasionally, a fewer partitions have fewer data points. In the worst case, the partition may be empty. One approach to overcome the above limitation is to re-partitioning a partition into two or more partitions. It is important to note that when data density in a partition is non-uniform, we re-partition if for having sub-partitions of near uniform data cardinality. The work by Gangwar et al. [105] is an attempt to achieve re-partitioning of a time-series for efficient prediction.

Most of the works on fuzzy time series presumed time invariant model of the time-series. The work by Aladag et al. [102] proposes a novel time-variant model fuzzy time-series, where they use particle swarm optimization techniques to determine fuzzy relational matrices connecting the membership functions of the of the two consecutive time point values. Besides providing a new approach to construct fuzzy relational matrices, the work also considers employing fuzzy c-means clustering techniques for fuzzification of the time-series.

The work introduced by Singh et al. [98] proposes four new issues, including (i) determination of effective length of intervals, (ii) handling of fuzzy logical relationship, (iii) determination of weight for each logical relationship and (iv) defuzzification of fuzzified time- series values. To handle the first problem, a new defuzzification technique of fuzzy time-series is introduced. Next fuzzy logical relationships are constructed from the fuzzified time series. Unlike most of the existing literature, that ignores multiple occurrences of repeated fuzzy logical relationships, here we use the multiple occurrences as higher weights to a fuzzy logical relation. These weights are later used in the defuzzification technique to obtain actual time series value. The proposed method validated with standard time series data and compared with existing techniques outperforms all the existing techniques by a large margin with respect to suitably selected metrics.

Fuzzy time series is currently receiving attention for its increasing application in business forecasting. Existing methods employ fuzzy clustering and genetic algorithm for fuzzification and artificial neural networks for determining relationship between successive data points in the time series. The work cited in [102] by Egrioglu et al. introduced a hybrid fuzzy time series approach to reach more accurate forecasts. Here the authors used fuzzy c-means clustering and artificial neural network for fuzzification and defining fuzzy relationship respectively. They employed proposed technique in enrollment data of University of Alabama. Huarng et al. in [103] presented a novel algorithm for automatic threshold selection on Secondary Factor for selection of rules for firing. The main points of their algorithm are given below.

1. Let $c(t)$ be the close price of secondary factor of a time series.
2. we take $v(t) = \frac{c(t)-c(t-1)}{t-(t-1)} = c(t) - c(t-1)$ as a variation series.
3. Take $|v(t)|$ and .... Threshold by the following steps

   (a) Compute average value of $|v(t)|$ by

$$v_{av} = \frac{\sum_{k=1}^{k-1} |v(t)|}{(k-1)} \tag{1.23}$$

   (b) Threshold $Th = \lfloor \frac{v_{av}}{B} + 1 \rfloor \times B$ where B is base factor arbitrarily taken as 10 for NASDAQ and DOW JONES (secondary factors) and $\lfloor x \rfloor$ is the lower ceiling value of $x$.

4. The selection criteria of the consequent fuzzy sets is dependent on the sign of $v_D$ and $v_N$. We first need to order the fuzzy sets in the consequent of the selecting for firing. Let the consequent fuzzy sets of the rule $A_j, A_k, A_l$. Then we place them in ascending order of the center of propositions, i.e., $A_j, A_k, A_l$. If $c(P_k) < c(P_j) < c(P_l)$, where $c(P_j)$ is the centered the close price of partition $P_j$.

Now if both the indices $v_D$ and $v_N$ are positive, then we select those fuzzy sets in the consequent where the corresponding partition centre value is $\geq$ the centre values of the partition corresponding to the antecedent fuzzy sets.

For example, if $c(P_k) < c(P_i)$ and $c(P_j) = c(P_i)$ and $c(P_l) < c(P_i)$ then we select $A_j$ and $A_l$ in the consequent and the corresponding rule is $A_i \rightarrow A_j,\ A_l$. In this manner we rectify the rules after they are selected for firing.

## 1.11 Scope of the Book

The book aims at designing novel techniques for economic time-series prediction. Because of unknown external parametric variations, the economic time-series exhibits an apparently random fluctuation, thereby making the prediction hard. The logic of fuzzy sets has proved itself a successful tool for uncertainty management in real-world systems. This motivated us to use the logic of fuzzy sets to utilize its fullest power of uncertainty management. One approach to handle the problem is to employ type-2 fuzzy sets that can capture both intra- and inter-personal level uncertainty. Among the two well-known variants of type-2 fuzzy sets, Interval Type-2 Fuzzy Set (IT2FS) is currently gaining popularity primarily for its simplicity and low computational complexity. This inspired us to model uncertainty in time-series using interval type-2 fuzzy sets. The IT2FS modeling is performed by constructing a set of prediction rules considering the current and the next sample points of the time-series. Such first-order dependence relations are widely used in the literature as the previous values (prior to current time point) of the time-series usually cannot model the temporal fluctuations in the series due to stray external disturbances that might occur between the current and the next sample points.

Besides IT2FS rule based models, another alternative approach adopted in the book is introduction of machine learning techniques for time-series prediction. Existing works model dynamic behavior of a time-series by supervised learning, which looks like representing the rules of prediction by supervised learning, such as pre-trained feed-forward neural networks. Supervised learning offers good prediction results when the fluctuation in the time-series does not depend largely on external parameters. Alternatively, unsupervised models cluster vertically chopped time-series components and describe them by a cluster centre. Naturally, a large daily time-series of 20 years having 365 × 20 sample points, can be fragmented into fewer clusters, each with a cluster centre that too describe a small duration time-series. Prediction at a given time point here refers to determining a suitable cluster centre describing a small fragment of the time-series that should appear next. In this book, we would consider both supervised and unsupervised approach to machine learning for time-series prediction.

The book includes five chapters. This chapter reviews the existing works on time-series prediction with a primary focus on fuzzy and machine learning approaches to time-series prediction. Chapter 2 provides a novel approach to time-series prediction by type-2 fuzzy logic. Here, the current to next point transitions are represented as type-2 production rules. Naturally, reasoning involved for prediction includes type-2 fuzzy logic. In particular, we here employed IT2FS model with provisions for auto-tuning of membership functions. Such tuning helps learning the current changes in the time-series dynamics.

In Chap. 3, we present a new formulation of hybrid type-1, type-2 reasoning, particularly considering in mind that we always cannot represent a partition by a type-2 membership function. When there exist quite a few distributed fragments of the series in a given partition, we model the partition by an interval type-2 fuzzy set. On the other hand, when there exist only a few data points forming a contiguous region, we model the partition by a type-1 fuzzy set. Now, when a rule contains both type-1 and type-2 fuzzy sets either in the antecedent or the consequent, we use hybrid fuzzy reasoning introduced in this chapter.

Chapter 4 proposes a clustering based approach, where the homogeneous fragments of the time-series, called segments are grouped into clusters, and the cluster centers, representing selective structures of the time-series are used for prediction of the next structural pattern to emerge at the next day. The encoding of transition of the time-series from one partition to the other following selected structures is performed with the help of a specialized automaton, called dynamic stochastic automaton. The merit of the chapter lies in the prediction of structures along with its expected duration and probability of occurrence, which has good impact on financial projections.

Chapter 5 proposes a novel approach for time-series prediction using neural nets. This approach is different from existing neural techniques for prediction by the following counts. It is capable of firing a set of concurrently firable prediction rules, realized with different neural nets. The prediction is performed by considering weighted sum of the inferences generated from all the fired prediction rules.

The last chapter is the concluding chapter of the book. It examines the scope of the work in view of the results obtained and performance metrics. It also explores the scope of future research in the present direction.

## 1.12 Summary

The chapter attempts to provide a foundation of time-series and its prediction mechanisms. The fundamental loophole in time-series prediction lies in the randomness of the series, which ultimately introduces uncertainty in the prediction algorithms. The chapter begins with possible sources of randomness in a time-series, including nonlinearity, non-Gaussian-ness, non-determinism and non-stationary characteristics of the time-series. It also demonstrates the scope of machine learning and fuzzy sets to address the prediction problem of a time-series in presence of uncertainty. The book would address two important problems of time-series prediction. The first problem refers to next time-point value prediction under uncertainty and non-determinism of prediction rules. The second problem is to predict the structure, comprising next few consecutive data points, using machine learning techniques. Although several variants of traditional machine learning algorithms could be attempted to handle the problem, we solved the problem using a clustering approach. The chapter ends with a discussion on the scope of the book.

**Exercises**

1. Check whether the data points given below follow the Gaussian distribution.

| x | 1 | 2 | 6 | 8 | 9 | 10 | 11 | 12 |
|---|---|---|---|---|---|---|---|---|
| f(x) | 3 | 4 | 7 | 660 | 11 | 12 | 6 | 7 |

[**Hints:** Obtain $\bar{f}(x) = \sum_{x=1}^{12} f(x)/12 \, and \, \sigma^2 = \frac{\sum_{x=1}^{12} (f(x) - \bar{f}(x))}{12}$. Then check whether $f(x)$ lies in $[\bar{f}(x) - 3\sigma, \bar{f}(x) + 3\sigma]$ for all x in [1, 12].]

2. Clarify the set of data points: $f(x)$ versus $x$ into two classes using a straight line.

| x | 2 | 3 | 4 | 6 | 8 | 10 | 12 |
|---|---|---|---|---|---|---|---|
| f(x) | 1 | 2 | 3 | 4 | 60 | 72 | 66 |

[**Hints:** Draw a straight line classifier that would segregate the points below (6, 4) and points above (8, 60) by this line. Hence determine the slope and y-intersect.]

3. Show that fuzzy implication relation of Lukasiewicz type is always greater or equal to the implication relation by Deins-Rescher implication function.

   [**Hints:** Deins-Rescher implication is given by,

$$\begin{aligned} R(x,y) &= \mathrm{Max}[\mu_{\bar{A}}(x_i), \mu_B(y_i)] \\ &= \mathrm{Max}[1 - \mu_A(x), \mu_B(y)] \\ &= \mathrm{Min}[1, (1 - \mu_A(x) + \mu_B(y))] \\ &= R(x,y) \text{ by } \textit{Lukasiewicz} \text{ implification.]} \end{aligned}$$

4. Let $A_i, A_j, A_k, A_l$ be four partitions of a time series. Given the prediction rules,

$$A_i \to A_j \text{ with } \mathrm{P(A_j/A_i)} = 0.3$$
$$A_i \to A_k \text{ with } \mathrm{P(A_k/A_i)} = 0.1$$
$$A_i \to A_l \text{ with } \mathrm{P(A_l/A_i)} = 0.5$$

Let P(that the time-series at time t falls in $A_i$) = 0.8 and the mid ranges of $A_j$, $A_k$ and $A_l$ be 20, 16, 14 K respectively, determine the expected value of series at time $t+1$.

[**Hints:** The probability of transitions to $A_j$, $A_k$, $A_l$ is given by 0.8 × 0.3 = 0.24, 0.8 × 0.2 = 0.16, 0.8 × 0.5 = 0.40. The expected value of the series at time $(t+1)$ is (0.24 × 20 K + 0.16 × 16 k + 0.40 × 14 K)/(0.24 + 0.16 + 0.40) = 16.2 K.]

5. Let R be the implication relation for transition from partition $P_i$ to $P_j$ given by,

| To $P_j$ / From $P_i$ | $P_1$ | $P_2$ | $P_3$ |
|---|---|---|---|
| $P_1$ | 0.6 | 0.2 | 0.2 |
| $P_2$ | 0.3 | 0.4 | 0.3 |
| $P_3$ | 0.2 | 0.5 | 0.3 |

Suppose, the close price c(t − 1) falls in partition $P_1$ with membership 0.6 and in $P_2$ with membership 0.2, and in $P_3$ with membership 0.2, compute $[\mu_{P_1}(c(t)\ \mu_{P_2}(c(t)\ \mu_{P_3}(c(t)]$.

[**Hints:**

$$[\mu_{P_1}(c(t)\ \mu_{P_2}(c(t)\ \mu_{P_3}(c(t)]$$
$$\text{by } [\mu_{P_1}(c(t-1)\ \mu_{P_2}(c(t-1)\ \mu_{P_3}(c(t-1)] \circ R]$$

6. Determine the partition containing $c(t-1)$, given $R(t-1,\ t) = R(t,\ t+1) = R$ as given in problem 5.

[**Hints:**

$$\begin{aligned}&[\mu_{P_1}(c(t+1)\ \mu_{P_2}(c(t+1)\ \mu_{P_3}(c(t+1)]\\ &= [\mu_{P_1}(c(t-1)\ \mu_{P_2}(c(t-1)\ \mu_{P_3}(c(t-1)] \circ R(t-1,t) \circ R(t,\ t+1)\\ &= [\mu_{P_1}(c(t-1)\ \mu_{P_2}(c(t-1)\ \mu_{P_3}(c(t-1)] \circ R^2]\end{aligned}$$

7. Given the dynamic range of a time-series [10,000, 60,000]. Suppose we need to partition the series into 6 unequal parts, such that sum of the dynamic range remains within the said bounds. Let the partition width in the proposed partitioning scheme be 5000. If the partitions are set based on date-density, and the ratio of data density is $P_1 : P_2 : P_3 : P_4 : P_5 : P_6 = 1:2:1:3:2:1$, then determine the lower and upper bounds of the partition.

[**Hints:** Range of the 1st partition $= \frac{60{,}000-10{,}000}{1+2+1+3+2+1} \times 1 = 5000$ units.
Thus the beginning and end of 1st partition is [1000, 15,000]. Similarly, the range of the 2nd partition is $= \frac{60{,}000-10{,}000}{1+2+1+3+2+1} \times 2 = 10,000$ units. Thus the beginning and end of partition is [15,000, 25,000]. This process can be extended for other partitions as well.]

8. The stock price in a partition $P_i$ is approximated by a Gaussian membership with mean = 50,000 and standard deviation 2000. What is the probability that a stock price of 55,500 will fall in the partition $P_i$?

[**Hints:** Let x = 55,500, mean μ = 50,000, σ = 2000. Continuing a Gaussian distribution of the form,

$$= \frac{1}{\sigma\sqrt{2\pi}} e^{\frac{-(x-\mu)^2}{2\sigma^2}}$$

where,

$$\begin{aligned} P(x = 55,500) \\ &= \frac{1}{2000 \times \sqrt{2\pi}} e^{-\frac{(55,500-55,000)}{2\times(2000)^2}} \\ &= 0.12] \end{aligned}$$

## References


1. Hjorth, B. (1970). EEG analysis based on time domain properties. *Electroencephalography and Clinical Neurophysiology, 29,* 306–310.
2. Jansen, B. H., Bourne, J. R., & Ward, J. W. (1981). Autoregressive estimation of short segment spectra for computerized EEG analysis. *IEEE Transactions on Biomedical Engineering*, *28*(9).
3. Box, G. E. P., & Jenkins, G. M. (1970). *Time series analysis: Forecasting and control* (revised ed. 1976). San Francisco: 7 Holden Day.
4. Box, G. E. P., Jenkins, G. M., & Reinsel, G. C. (1994). *Time series analysis: Forecasting and control* (3rd ed.). Englewood Cliffs: NJ7 Prentice Hall.
5. Chatfield, C. (1988). What is the best method of forecasting? *Journal of Applied Statistics, 15,* 19–38.
6. Chevillon, G., & Hendry, D. F. (2005). Non-parametric direct multistep estimation for forecasting economic processes. *International Journal of Forecasting, 21,* 201–218.
7. Cholette, P. A. (1982). Prior information and ARIMA forecasting. *Journal of Forecasting, 1,* 375–383.
8. De Alba, E. (1993). Constrained forecasting in autoregressive time series models: A Bayesian analysis. *International Journal of Forecasting, 9,* 95–108.
9. Arin˜o, M. A., & Franses, P. H. (2000). Forecasting the levels of vector autoregressive log-transformed time series. *International Journal of Forecasting*, *16*, 111–116.
10. Artis, M. J., & Zhang, W. (1990). BVAR forecasts for the G-7. *International Journal of Forecasting, 6,* 349–362.
11. Ashley, R. (1988). On the relative worth of recent macroeconomic forecasts. *International Journal of Forecasting, 4,* 363–376.
12. Bhansali, R. J. (1996). Asymptotically efficient autoregressive model selection for multistep prediction. *Annals of the Institute of Statistical Mathematics, 48,* 577–602.
13. Bhansali, R. J. (1999). Autoregressive model selection for multistep prediction. *Journal of Statistical Planning and Inference*, *78*, 295–305.

14. Bianchi, L., Jarrett, J., & Hanumara, T. C. (1998). Improving forecasting for telemarketing centers by ARIMA modelling with interventions. *International Journal of Forecasting, 14,* 497–504.
15. Bidarkota, P. V. (1998). The comparative forecast performance of univariate and multivariate models: An application to real interest rate forecasting. *International Journal of Forecasting, 14,* 457–468.
16. Scarborough, J. (1983). *Numerical Mathematical Analysis*, Oxford.
17. Konar, A. (2006). *Computational intelligence: Principles*. Techniques and Applications: Springer.
18. Soman, K. P., Loganathan, R., & Ajay, V. (2009). *Machine learning using SVM and other kernel methods*. India: Prentice-Hall.
19. Dua, H., & Zhanga, N. (2008). Time series prediction using evolving radial basis function networks with new encoding scheme. *Elsevier Neurocomputing, 71,* 1388–1400.
20. Frank, R. J., Davey, N., & Hunt, S. P. (2001). Time series prediction and neural networks. *Journal of Intelligent and Robotic Systems, 31*(1–3), 91–103.
21. Faraway, J., & Chatfield, C. (1998). Time series forecasting with neural networks: A comparative study using the air line data. *Journal of the Royal Statistical Society: Series C (Applied Statistics), 47,* 231–250.
22. Meesad, P., & Rase, R. I. (2013). Predicting stock market price using support vector regression. In *IEEE conference,* Dhaka (pp. 1–6). May 17–18, 2013.
23. Lai, L. K. C., & Liu J. N. K. (2010). Stock forecasting using support vector machine. In *Proceedings of the Ninth International Conference on Machine Learning and Cybernetics* (pp. 1607–1614).
24. Debasish, B., Srimanta, P., & Dipak, C. P. (2007). Support vector regression. *Neural Information Processing—Letters and Reviews, 11*(10), 203–224.
25. Li, X., & Deng, Z. (2007). A machine learning approach to predict turning points for chaotic financial time series. In *19th IEEE International Conference on Tools with Artificial Intelligence.*
26. Kattan, A., Fatima, S., & Arif, M. (2015). Time-series event-based prediction: An Unsupervised learning framework based on genetic programming. *Elsevier Information Sciences, 301,* 99–123.
27. Koza, J. R. (1992). *Genetic Programming*. Cambridge: MIT Press.
28. Miranian, A., & Abdollahzade, M. (2013). Developing a local least-squares support vector machines-based neuro-fuzzy model for nonlinear and chaotic time series prediction. *IEEE Transactions on Neural Networks and Learning Systems, 24*(2), 207–218.
29. Yan, W. (2012). Toward automatic time-series forecasting using neural networks. *IEEE Transactions on Neural Networks and Learning Systems, 23*(7), 1028–1039.
30. Lee, C. L., Liu, A., & Sung, W. (2006). Pattern discovery of fuzzy time series for financial prediction. *IEEE Transactions on Knowledge and Data Engineering, 18*(5), 61325.
31. Coyle, D., Prasad, G., & McGinnity, T. M. (2005, December). A time-series prediction approach for feature extraction in a brain–computer interface. *IEEE Transactions on Neural Systems and Rehabilitation Engineering, 13*(4), 461–467.
32. Dash, R., Dash, P. K., & Bisoi, R. (2014). A self adaptive differential harmony search based optimized extreme learning machine for financial time-series prediction. *Swarm and Evolutionary Computation, 19,* 25–42.
33. Batres-Estrada, G. (2015). Deep learning for multivariate financial time series. http://www.math.kth.se/matstat/seminarier/reports/M-exjobb15/150612a.pdf. June 4, 2015.
34. Hrasko, R., Pacheco, A. G. C., & Krohling, R. A. (2015). Time series prediction using restricted boltzmann machines backpropagation. *Elsevies, Procedia Computer Science, 55,* 990–999.
35. Zimmermann, H.-J. (2001). *Fuzzy set theory and its applications*. New York: Springer Science.
36. Ross, T. J. (2004). *Fuzzy logic with engineering applications*, Chichester: Wiley IEEE press.

37. Song, Q., & Chissom, B. S. (1993). Fuzzy time series and its model. *Fuzzy Sets and Systems, 54*(3), 269–277.
38. Song, Q., & Chissom, B. S. (1993). Forecasting enrollments with fuzzy time series—Part I. *Fuzzy Sets and Systems, 54*(1), 1–9.
39. Song, Q., & Chissom, B. S. (1994). Forecasting enrollments with fuzzy time series—Part II. *Fuzzy Sets and Systems, 62*(1), 1–8.
40. Song, Q. (2003). A note on fuzzy time series model selection with sample autocorrelation functions. *Cybern. Syst., 34*(2), 93–107.
41. Chen, S. M., Chu, H. P., Sheu, T. W. (2012, November). TAIEX forecasting vusing fuzzy time series and automatically generated weights of multiple factors. *IEEE Transactions on Systems, Man, and Cybernetics-Part A: Systems and Humans, 42*(6) 1485–1495.
42. Chen, S. M. (1996). Forecasting enrollments based on fuzzy time series. *Fuzzy Sets and Systems, 81*(3), 311–319.
43. Chen, S. M., & Chang, Y. C. (2010). Multi-variable fuzzy forecasting based on fuzzy clustering and fuzzy rule interpolation techniques. *Information Sciences, 180*(24), 4772–4783.
44. Chen, S. M., & Chen, C. D. (2011). Handling forecasting problems based on high-order fuzzy logical relationships. *Expert Systems with Applications, 38*(4), 3857–3864.
45. Chen, S. M., & Hwang, J. R. (2000). Temperature prediction using fuzzy time series. *IEEE Transactions on Systems, Man, and Cybernetics. Part B, Cybernetics, 30*(2), 263–275.
46. Chen, S. M., & Tanuwijaya, K. (2011). Multivariate fuzzy forecasting based on fuzzy time series and automatic clustering clustering techniques. *Expert Systems with Applications, 38*(8), 10594–10605.
47. Chen, S. M., & Tanuwijaya, K. (2011). Fuzzy forecasting based on high-order fuzzy logical relationships and automatic clustering techniques. *Expert Systems with Applications, 38*(12), 15425–15437.
48. Chen, S. M., & Wang, N. Y. (2010). Fuzzy forecasting based on fuzzy-trend logical relationship groups. *IEEE Transactions on Systems, Man, and Cybernetics. Part B, Cybernetics, 40*(5), 1343–1358.
49. Chen, S. M., Wang, N. Y., & Pan, J. S. (2009). Forecasting enrollments using automatic clustering techniques and fuzzy logical relationships. *Expert Systems with Applications, 36*(8), 11070–11076.
50. Chen, S. M., & Chen, C. D. (2011). TAIEX forecasting based on fuzzy time series and fuzzy variation groups. *IEEE Transactions on Fuzzy Systems, 19*(1), 1–12.
51. Lu, W., Yang, J., Liu, X., & Pedrycz, W. (2014). The modeling and prediction of time-series based on synergy of high order fuzzy cognitive map and fuzzy c-means clustering. *Elsevier, Knowledge Based Systems, 70,* 242–255.
52. Lee, L. W., Wang, L. H., & Chen, S. M. (2007). Temperature prediction and TAIFEX forecasting based on fuzzy logical relationships and genetic algorithms. *Expert Systems with Applications, 33*(3), 539–550.
53. Hwang, J. R., Chen, S. M., & Lee, C. H. (1998). Handling forecasting problems using fuzzy time series. *Fuzzy Sets and Systems, 100*(1–3), 217–228.
54. Yolcu, U., Egrioglu, E., Uslu, V. R., Basaran, M. A., & Aladag, C. H. (2009). A new approach for determining the length of intervals for fuzzy time series. *Applied Soft Computing, 9*(2), 647–651.
55. Yu, T. H. K., & Huarng, K. H. (2008). A bivariate fuzzy time series model to forecast the TAIEX. *Expert Systems with Applications, 34*(4), 2945–2952.
56. Yu, T. H. K., & Huarng, K. H. (2010). Corrigendum to "A bivariate fuzzy time series model to forecast the TAIEX". *Expert Systems with Applications, 37*(7), 5529.
57. Huarng, K., & Yu, H. K. (2005). A type 2 fuzzy time series model for stock index forecasting. *Physica A, 353,* 445–462.
58. Huarng, K., & Yu, T. H. K. (2006). The application of neural networks to forecast fuzzy time series. *Physica A, 363*(2), 481–491.

59. Huarng, K., Yu, H. K., & Hsu, Y. W. (2007). A multivariate heuristic model for fuzzy time-series forecasting. *IEEE Transactions on Systems, Man, and Cybernetics. Part B, Cybernetics, 37*(4), 836–846.
60. Yu, T. H. K., & Huarng, K. H. (2010). A neural network-based fuzzy time series model to improve forecasting. *Expert Systems with Applications, 37*(4), 3366–3372.
61. Huarng, K. (2001). Effective lengths of intervals to improve forecasting in fuzzy time-series. *Fuzzy Sets and Systems, 123,* 387–394.
62. Bhattacharya, D., Konar, A., & Das, P. (2016). Secondary factor induced stock index time-series prediction using self-adaptive interval type-2 fuzzy sets. *Neurocomputing, Elsevier, 171,* 551–568.
63. Jilani, T. A., Burney, S. M. A., & Ardil, C. (2010). Fuzzy metric approach for fuzzy time-series forecasting based on frequency density based partition. *World Academy of Science, Engineering and Technology, 4*(7), 1–5.
64. Sullivan, Joe, & Woodall, William H. (1994). A comparison of fuzzy forecasting and Markov modeling. *Fuzzy Sets and Systems, 64,* 279–293.
65. Kuo, I. H., Horng, S. J., Kao, T. W., Lin, T. L., Lee, C. L., & Pan, Y. (2009). An improved method for forecasting enrollments based on fuzzy time series and particle swarm optimization. *Expert Systems with Applications, 36*(3), 6108–6117.
66. Wong, H. L., Tu, Y. H., & Wang, C. C. (2009). An evaluation comparison between multivariate fuzzy time series with traditional time series model for forecasting Taiwan export. In *Proceedings of World Congress Computer Science and Information Engineering* (pp. 462–467).
67. Huarng, K., & Yu, T. H. (2012, October). Modeling fuzzy time series with multiple observations. *International Journal of Innovative, Computing, Information and Control, 8* (10(B)).
68. Caia, Q. S., Zhang, D., Wu, B., Leung, S. C. H. (2013). A novel stock forecasting model based on fuzzy time series and genetic algorithm. *Procedia Computer Science, 18,* 1155–1162.
69. Singh, P., & Borah, B. (2013). High-order fuzzy-neuro expert system for time series forecasting. *Knowledge-Based Systems, 46,* 12–21.
70. Kaushik, A., & Singh, A. K. (2013, July). Long term forecasting with fuzzy time series and neural network: A comparative study using Sugar production data. *International Journal of Computer Trends and Technology (IJCTT), 4*(7).
71. Uslu, V. R., Bas, E., Yolcu, U., & Egrioglu, E. (2014). A fuzzy time-series approach based on weights determined by the number of recurrences of fuzzy relations. *Swarm and Evolutionary Computation, 15,* 19–26.
72. Wong, W. K., Bai, E., & Chu, A. W. C. (2010, December). Adaptive time-variant models for fuzzy-time-series forecasting. *IEEE Transactions on Systems, Man, and Cybernetics—Part B: Cybernetics*, *40*(6), 1531–1542.
73. Saima H., Jaafar, J., Belhaouari, S., & Jillani, T. A. (2011, August). ARIMA based interval type-2 fuzzy model for forecasting. *International Journal of Computer Applications, 28*(3).
74. Tsaur, R., Yang, J., & Wang, H. (2005). Fuzzy relation analysis in fuzzy time series model. *Computers and Mathematics with Applications, 49*, 539–548.
75. Chen, S., & Shen, C. (2006). When wall street conflicts with main street—the divergent movements of Taiwan's leading indicators. *International Journal of Forecasting, 22,* 317–339.
76. Hong, D. H. (2005). A note on fuzzy time-series model. *Fuzzy Sets and Systems, 155,* 309–316.
77. Li, S. T., Cheng, Y. C., & Lin, S. Y. (2008). A FCM-based deterministic forecasting model for fuzzy time series. *Computers & Mathematics with Applications, 56,* 3052–3063.
78. Stevenson, M., & Porter, J. E. (2009). Fuzzy time series forecasting using percentage change as the universe of discourse. *World Academy of Science, Engineering and Technology, 3,* 138–141.

79. Chu, H. H., Chen, T. L., Cheng, C. H., & Huang, C. C. (2009). Fuzzy dual-factor time-series for stock index forecasting. *Expert Systems with Applications, 36,* 165–171.
80. Chena, T. L., Chenga, C. H., & Teoha, H. J. (2007). Fuzzy time-series based on Fibonacci sequence for stock price forecasting. *Physica A, 380,* 377–390.
81. Yu, H. K. (2005). Weighted fuzzy time series models for TAIEX forecasting. *Physica A, 349,* 609–624.
82. Lee, L. W., Wang, L. H., Chen, S. M., & Leu, Y. H. (2006). Handling forecasting problems based on two-factors high-order fuzzy time series. *IEEE Transactions on Fuzzy Systems, 14* (3), 468–477.
83. Lertworaprachaya, Y., Yang, Y., & John, R. (2010). High-order type-2 fuzzy time series. In *IEEE International Conference of Soft Computing and Pattern Recognition* (pp. 363–368).
84. Chena, T. L., Cheng, C. H., & Teoh, H. J. (2008). High-order fuzzy time-series based on multi-period adaptation model for forecasting stock markets. *Physica A, 387,* 876–888.
85. Hassan, S., Jaafar, J., Samir, B. B., & Jilani, T. A. (2012). A hybrid fuzzy time series model for forecasting. *Engineering Letters*, (*Advance online publication*: 27 February 2012).
86. Saxena, P., Sharma, K., & Easo, S. (2012). Forecasting enrollments based on fuzzy time series with higher forecast accuracy rate. *International Journal of Computer Technology & Applications, 3*(3), 957–961.
87. Saxena, P., & Easo, S. (2012). A new method for forecasting enrollments based on fuzzy time series with higher forecast accuracy rate. *International Journal of Computer Technology & Applications, 3*(6), 2033–2037.
88. Lee, C. H. L., Chen, W., & Liu, A. (2004). An implementation of knowledge based pattern recognition for financial prediction. In *Proceedings of the 2004 IEEE Conference on Cybernetics and Intelligent Systems*, Singapore, December 1–3, 2004.
89. Sadaei, H. J., & Lee, M. H. (2014). Multilayer stock forecasting model using fuzzy time series. *Hindawi Publishing Corporation, The Scientific World Journal, Article ID, 610594,* 1–10.
90. Yu, H. K. (2005). A refined fuzzy time-series model for forecasting. *Physica A, 346,* 657–681.
91. Singh, S. R. (2008). A computational method of forecasting based on fuzzy time series. *Mathematics and Computers in Simulation, 79,* 539–554.
92. Cheng, C. H., Chen, T. L., Teoh, H. J., & Chiang, C. H. (2008). Fuzzy time-series based on adaptive expectation model for TAIEX forecasting. *Expert Systems with Applications, 34,* 1126–1132.
93. Singh, S. R. (2007). A simple method of forecasting based on fuzzy time series. *Applied Mathematics and Computation, 186,* 330–339.
94. Bajestani, N. S., & Zare, A. (2011). Forecasting TAIEX using improved type 2 fuzzy time series. *Expert Systems with Applications, 38,* 5816–5821.
95. Huarng, K., & Yu, H. K. (2005). A Type 2 fuzzy time series model for stock index forecasting. *Physica A, 353,* 445–462.
96. Huarng, K., Yu, T., & Hsu, Y. W. (2007). A multivariate heuristic model for fuzzy time-series forecasting. *IEEE Transactions On Systems and Cybernatics, 37*(4), 836–846.
97. Mendel. J., Hagras, H., Tan, W., Melek, W. W. & Ying, H. (2014). Introduction to type-2 fuzzy logic control: Theory and applications. Wiley-IEEE Press.
98. Singh, P., & Borah, B. (2013). An efficient time series forecasting model based on fuzzy time series. *Elsevier Engineering Applications of Artificial Intelligence, 26,* 2443–2457.
99. Lu, W., Pedrycz, W., Liu, X., Yang, J., & Li, P. (2014). The modeling of time-series based on fuzzy information granules. *Expert Systems and Applications, 47,* 3799–3808.
100. Wang, W., Pedrycz, W., & Liu, X. (2015). Time-series long term forecasting model based on information granules and fuzzy clustering. *Engineering Application of Artificial Intelligence, 41,* 17–24.
101. Mendel, J. M., & Wu, D. (2004). *Perceptual Computing.* Wiley.

102. Aladag, C. H., Yolcu, U., Egrioglu, E., & Dalar, A. Z. (2012). A new time invariant fuzzy time series forecasting method based on particle swarm optimization. *Elsevier Applied Soft Computing, 12,* 3291–3299.
103. Huarng, K. (2001). Heuristic models of fuzzy time series for forecasting. *Fuzzy Sets and Systems, 123,* 369–386.
104. Li, S., & Cheng, Y. (2007). Deterministic fuzzy time series model for forecasting enrollments. *Elsevier Computers & Mathematics with Applications, 53*(12), 1904–1920.
105. Gangwar, S. S., & Kumar, S. (2012). Partitions based computational method for high-order fuzzy time series forecasting. *Elsevier Expert Systems with Applications, 39*, 12158–12164.

# Chapter 2
# Self-adaptive Interval Type-2 Fuzzy Set Induced Stock Index Prediction

**Abstract** This chapter introduces an alternative approach to time-series prediction for stock index data using Interval Type-2 Fuzzy Sets. The work differs from the existing research on time-series prediction by the following counts. First, partitions of the time-series, obtained by fragmenting its valuation space over disjoint equal sized intervals, are represented by Interval Type-2 Fuzzy Sets (or Type-1 fuzzy sets in absence of sufficient data points in the partitions). Second, an interval type-2 (or type-1) fuzzy reasoning is performed using prediction rules, extracted from the (main factor) time-series. Third, a type-2 (or type-1) centroidal defuzzification is undertaken to determine crisp measure of inferences obtained from the fired rules, and lastly a weighted averaging of the defuzzified outcomes of the fired rules is performed to predict the time-series at the next time point from its current value. Besides the above three main prediction steps, the other issues considered in this chapter include: (i) employing a new strategy to induce the main factor time-series prediction by its secondary factors (other reference time-series), and (ii) self-adaptation of membership functions to properly tune them to capture the sudden changes in the main-factor time-series. Performance analysis undertaken reveals that the proposed prediction algorithm outperforms existing algorithms with respect to root mean-square error by a large margin ($\geq 23\%$). A statistical analysis undertaken with paired t-test confirms that the proposed method is superior in performance at 95% confidence level to most of the existing techniques with root mean square error as the key metric.

### Abbreviations

| | |
|---|---|
| CSV | Composite secondary variation |
| CSVS | Composite secondary variation series |
| FOU | Footprint of uncertainty |
| IT2 | Interval type-2 |
| IT2FS | Interval type-2 fuzzy set |
| IT2MF | Interval type-2 membership function |
| MFCP | Main factor close price |
| MFTS | Main factor time-series |
| MFVS | Main factor variation series |



A. Konar and D. Bhattacharya, *Time-Series Prediction and Applications*,
Intelligent Systems Reference Library 127,
DOI 10.1007/978-3-319-54597-4_2

RMSE Root mean square error
SFTS Secondary factor time-series
SFVS Secondary factor variation series
SFVTS Secondary factor variation time-series
T1 Type-1
T1FS Type-1 fuzzy set
VTS Variation time-series

**Symbols**

$A_{i,j}$ Type-1 fuzzy set for partition $P_i$ of MFTS
$\tilde{A}_i$ Interval type-2 fuzzy set for partition $P_i$ in MFTS
$B_i$ Classical set for partition $Q_i$ of MFVS
$B'_i$ Classical set for partition $Q_i$ for SFVS/CSVS
$c(t)$ Close Price on $t$th day
$c_l$ Left end point centroid of IT2FS
$c_r$ Right end point centroid of IT2FS
$c$ Centroid of an IT2FS
$c'$ Measured value of $c(t)$ in centroid calculation
$c^x$ Type-1 centroid of $\mu_{A_x}(c(t))$
$m_A$ Mean values of the distributions of RMSE obtained by algorithms A
$P_i$ *I*th partition for close price time series (of MFTS)
$Q_i$ *I*th partition for variation series (of MFVS/SFVS/CSVS)
$s_A$ Standard deviation of the respective samples obtained by algorithms A
$V_M^d(t)$ Main factor variation series with delay $d$
$V_{S^i}^d$ *I*th secondary factor variation series with delay $d$
$V_S^d(t)$ Composite secondary variation series with delay $d$
$W_{S^i}$ Weight for $i$th secondary factor
$\mu_A(x)$ Type-1 membership function of linguistic variable $x$ in fuzzy set $A$
$\bar{\mu}_{\tilde{A}}(x)$ Upper membership function of IT2FS $\tilde{A}$
$\underline{\mu}_{\tilde{A}}(x)$ Lower membership function of IT2FS $\tilde{A}$
$\Delta_{S^i}$ Total difference variation for CSVS of $i$th secondary factor
$\hat{\Delta}_{S^i}$ Normalized value of $\Delta_{S^i}$ for CSVS of $i$th secondary factor

## 2.1 Introduction

Prediction of a time-series [1] refers to determining the amplitude of the series at time $t + 1$ from its previous m sample values located at time: $t, t - 1, t - 2, \ldots, t - (m - 1)$ for a finite positive integer $m$. An $m$-th order time-series prediction involves all the m previous sample values directly for its forecasting/prediction [2, 3]. In this chapter, we, for the sake of simplicity, however, use a first order prediction of time-series, where the $(t + 1)$-th sample of the time-series directly depends only on the sample value at time $(t + 1 - d)$, where $d$ denotes the time-delay, although all the previous m sample values are required to design the prediction rules. There exists a vast literature on prediction of time-series for real processes, including rainfall [4, 5], population growth [6], atmospheric temperature [7], university enrollment for students [8–11], economic growth [12] and the like. This chapter is concerned with stock index, the time-series of which describing close price [13], is characterized by the following four attributes: non-linear [14], non-deterministic, non-stationary [15] and non-Gaussian jointly.

Designing a suitable model for stock index prediction requires handling the above four characteristics jointly. Although there exist several attempts to model time-series using non-linear oscillators [16], non-linear regression [17], adaptive auto-regression [18], Hzorth parameters [19] and the like, none of these could accurately model these time-series [20] for their inherent limitations to capture all the four characteristics jointly.

The logic of fuzzy sets plays a promising role to handle the above problems jointly. First, the nonlinearity of time-series is modeled by the nonlinearity of membership functions and their nonlinear mapping from antecedent to consequent space of fuzzy production rules. Second, the non-deterministic characteristics of the time-series (that might occur due to randomness in a wide space), is here significantly reduced because of its occurrence in one of a few equal sized partitions of the universe of discourse. Third, the non-stationary characteristics of the time-series that offers a correlation of signal frequencies with time [15] is avoided in fuzzy modeling by time-invariant models of membership functions [8]. Lastly, the non-Gaussian feature may be relaxed as locally Gaussian within small intervals (partitions) of the time-series. Thus, fuzzy sets are capable of capturing the uncertainty/imprecision in time-series prediction that might arise because of the above four hindrances.

The inherent power of fuzzy sets to model uncertainty of time-series has attracted researchers to employ fuzzy logic in time-series prediction. Song et al. [8–11] pioneered the art of fuzzy time-series prediction by representing the time-series value at time $t - 1$ and time $t$ as fuzzy membership functions (MFs) and connected them by fuzzy implication relations for all possible time t in the time-series. If there exist n possible discrete values of time $t$, then we would have $n - 1$ possible fuzzy implication relations. Song et al. combined all these implication relations into a single relation R by taking union of all of these relations. The prediction involves first fuzzifying the crisp value of the time series at time $t$ and then using composition rule of inference to determine the MF of the predicted time series at time $t + 1$ using R as the composite time-invariant implication relation. Lastly, they defuzzified the result to obtain the crisp value of the time-series at time $t + 1$.

The fundamental deviation in the subsequent work by Chen [21] lies in grouping of rules having common antecedents. Thus during the prediction phase, only few rules whose antecedent match with the antecedent of the fuzzified time-series value at time $t$ only, need to be fired to obtain multiple MFs of the inferred consequences, one for each fired rule, an averaging type of defuzzification of which yields the actual inference at time $t + 1$. Hwang et al. considered a variation time-series [22] by taking the difference of two consecutive values of the time-series, and used max-product compositional rule of inference to predict the inference of the variation at time $t + 1$ from its previous values. A weighted average type of defuzzification was used to obtain the predicted value of the time-series at time $t + 1$. Cai et al. [23] introduced genetic algorithm to determine the optimal weight matrix for transitions of partitions of a given time-series from each day to its next day, and used the weight matrix to predict the time-series at time $t + 1$ from its value at time $t$. In [7], Chen et al. extended the work of Hwang et al. by first introducing a concept of secondary factors in the prediction of main factor time-series. There exists a vast literature on time-series prediction using fuzzy logic. A few of these that deserve special mention includes adaptive time-variant modeling [24], adaptive expectation modeling [25], Fibonacci sequence [26], Neural networks [27, 28], Particle Swarm Optimization [29] based modeling, fuzzy cognitive maps and fuzzy clustering [30], bi-variate [31, 32] and multi-variate [33–37] modeling and High order fuzzy multi-period adaptation model [38] for time-series prediction.

Most of the traditional works on stock index prediction developed with fuzzy logic [39] employ type-1(T1) fuzzy reasoning to predict future stock indices. Although T1 fuzzy sets have proved their excellence in automated reasoning for problems of diverse domains, including fuzzy washing machines [40, 41], fuzzy color TV [42] etc., they have limited power to capture the uncertainty of the real world problems [43]. Naturally, T1 fuzzy logic is incompetent to stock (and general time-series) prediction problems. The importance of interval type-2 fuzzy set (IT2FS) over its type-1 counterpart in chaotic time-series prediction has already been demonstrated by Karnik and Mendel [44]. There exist a few recent works attempting to model stock prediction problem using type-2 fuzzy sets [45, 46]. These models aimat representing a single (interval) type-2 membership function (MF), considering three distinct stock data items, called close, high and low prices [13]. Here too, the authors partitioned each of the above three time-series into intervals of equal size, and represented each partition as T1 fuzzy set. They constructed fuzzy If-Then rules describing transitions of stock index price from one day to the next day for each of the above time series. During prediction, they identified a set of rules containing antecedent fuzzy sets corresponding to current stock prices, obtained union and intersection of the consequents of the rules to derive (interval) type-2 fuzzy inferences and employed centre average defuzzifiers to predict the stock price for the next day. Bagestani and Zare [46] extended the above work by adaptation of the structure of the membership functions and weights of the defuzzified outputs to optimize root mean square error. In addition, the latter work employed centre of gravity defuzzifier in place of centre average defuzzifier used previously. The present chapter is an extension of the seminal work of Chen et al. [47] by the following counts.

1. In order to represent the close price $c(t)$ within a partition (interval), we represent each short duration contiguous fluctuation of $c(t)$ in a given partition of the universe of $c(t)$ by a type-1 MF, and take union of all these type-1 MFs within a partition to represent it by an interval type-2 fuzzy set (IT2FS). Under special circumstances, when a partition includes one or a few contiguous data points only, we represent the partition by a type-1 MF only.
2. The antecedent and consequent of fuzzy prediction rules of the form $A_i \rightarrow A_j$ (extracted from the consecutive occurrence of data points in partitions $P_i$ and $P_j$, are represented by interval type-2 (IT2) (or type-1) fuzzy sets depending on the count and consecutive occurrences of data points in a partition. Naturally, there exist four possible types of fuzzy prediction rules: IT2 to IT2, IT2 to type-1, type-1 to IT2 and type-1 to type-1 depending on the representation of $A_i$ and $A_j$ by IT2 or type-1 fuzzy sets. This chapter thus employs four possible types of reasoning, each one for one specific type of prediction rule.
3. Appropriate defuzzification techniques, such as Karnik-Mendel algorithm for IT2 inferences [48, 49] and centroidal defuzzification for type-1 inferences have been employed to obtain the predicted close price at day $t + 1$.
4. Existing works [47] presume that the variation in secondary factor of the current day (of reliable reference time-series) identically influences the main factor of the next day. Naturally, if the interval counts in both the secondary factor and the main factor are equal, then the above variations have the same interval label in their respective universes. In the present chapter, we relax the restriction by considering all possible occurrence of variation of the main factor intervals for each occurrence of secondary factor interval obtained from the historical data. Such relaxation keeps the prediction process free from bias. In case, the current occurrence of secondary factor interval has no precedence in the historical variation data, we adopt the same principle used in [47].
5. One additional feature that caused significant improvement in performance in prediction is due to the introduction of evolutionary adaptation in the parameters of the selected structure of membership functions. The evolutionary adaptation tunes the base-width of the triangular/Gaussian membership functions (MFs) employed to reduce the root mean square error (RMSE) [27, 36]. Experiments undertaken reveal that tuning of parameters of MFs result in over 15% improvement in RMSE.

The proposed extensions introduced above outperforms all existing works on IT2 [45, 46] and type-1 fuzzy logic based stock index prediction techniques [8–11, 21, 47, 50] using RMSE as the metric.

The rest of this chapter is divided into five sections. In Sect. 2.2, we provide the necessary definitions required to understand this chapter. In Sect. 2.3, we present both training and prediction algorithms using IT2 fuzzy reasoning in the context of stock price prediction. Section 2.4 is concerned with experimental issues and computer simulation with details of results obtained and their interpretation. Performance analysis of the proposed technique with existing works is compared in Sect. 2.5. Conclusions are listed in Sect. 2.6.

## 2.2 Preliminaries

This section provides a few fundamental definitions pertaining to both time-series prediction and IT2FS. These definitions will be used in the rest of this chapter.

**Definition 2.1** The last traded price in a trading day of a stock index is called *close price*, hereafter denoted by $c(t)$.

**Definition 2.2** The stock index under consideration for prediction of a time series is called *Main Factor Time Series* (MFTS). Here, we consider TAIEX (Taiwan Stock Exchange Index) as the MFTS.

**Definition 2.3** The associated indices of time series that largely influence prediction of the MFTS is called *Secondary Factor Time Series* (SFTS). Here, we consider NASDAQ (National Association of Securities and Dealers Automated Quotations) and DJI (Dow Jones Industrial average) as the SFTS.

**Definition 2.4** For a given close price time series (CTS) $c(t)$, the *Variation Time Series* (VTS) [47] *with delay of d days* for close price is given by,

$$VTS^{d}(t) = \frac{c(t) - c(t-d)}{c(t-d)} \times 100 \tag{2.1}$$

for $t \in [t_{\min}, t_{\max}]$, where $t_{min}$ and $t_{max}$ denote the beginning and terminating days of the *training period* [47]. Here we consider $V_M^d(t)$ and $V_S^d(t)$ as the VTS for MFTS and SFTS respectively.

**Definition 2.5** Prediction of MFTS $c(t+d)$ here refers to determining $c(t+d)$ from its historical values: $c(t)$, $c(t-1)$, $c(t-2)$, $c(t-3)$, …, $c(t-(m-1))$ and secondary factor VTS (SFVTS) $V_S^d(t)$, $V_S^d(t-1)$, …, $V_S^d(t-(m-1))$ for some positive integer m. Such prediction is referred to as m-th order forecasting. However, in most of the applications, researchers take $d = 1$ for simplicity and convenience [8–11, 47].

**Definition 2.6** A T1 fuzzy set is a two tuple given by $<\mathrm{x}, \mu_{\mathrm{A}}(\mathrm{x})>$ where x is a linguistic variable in a Universe X and $\mu_{\mathrm{A}}(\mathrm{x})$ is the membership function of $x$ in fuzzy set $A$, where $\mu_A(x) \in [0, 1]$.

**Definition 2.7** A general type-2 Fuzzy Set (GT2FS) is a three tuple given by $<\mathrm{x}, \mu_{\mathrm{A}}(\mathrm{x}), \mu_{\mathrm{A}}(\mathrm{x}, \mu_{\mathrm{A}}(\mathrm{x}))>$ where $x$ and $\mu_{\mathrm{A}}(\mathrm{x})$ have the same meaning as in Definition 2.6, and $\mu(\mathrm{x}, \mu_{\mathrm{A}}(\mathrm{x}))$ is the secondary membership in [0, 1] at a given $(x, \mu_{\mathrm{A}}(\mathrm{x}))$.

**Definition 2.8** An interval type-2 fuzzy set (IT2FS) is defined by two T1 membership functions (MFs), called *Upper Membership Function (UMF)*, and *Lower Membership Function (LMF)*. An IT2FS $\tilde{A}$, therefore, is represented by $<\underline{\mu}_{\tilde{A}}(x), \bar{\mu}_{\tilde{A}}(x)>$ where $\underline{\mu}_{\tilde{A}}(x)$ and $\bar{\mu}_{\tilde{A}}(x)$ denote the lower and upper membership functions respectively. The secondary membership $\mu(x,\mu_A(x))$ in IT2FS is considered as 1 for all $x$ and $\mu_A(x)$.

**Definition 2.9** The left end point centroid is the smallest of all possible centroids (of the embedded fuzzy sets [48]) in an IT2FS $\tilde{A}$ and is evaluated by

$$c_l = \frac{\sum_{i=1}^{k-1} \bar{\mu}_{\tilde{A}}(x_i) \cdot x_i + \sum_{i=k+1}^{N} \underline{\mu}_{\tilde{A}}(x_i) \cdot x_i}{\sum_{i=1}^{k-1} \bar{\mu}_{\tilde{A}}(x_i) + \sum_{i=k+1}^{N} \underline{\mu}_{\tilde{A}}(x_i)} \tag{2.2}$$

using the well-known Karnik-Mendel algorithm [51],

where $x \in \{x_1, x_2, \ldots, x_N\}$ and $x_{i+1} > x_i \ \forall i = 1 \text{ to } N-1$. Here $x = x_k$ is a switch point and $N$ denotes the number of sample points of $\bar{\mu}_{\tilde{A}}(x_i)$ and $\underline{\mu}_{\tilde{A}}(x_i)$.

**Definition 2.10** The right end point centroid is the largest of all possible centroids (of the embedded fuzzy sets [48]) in an IT2FS $\tilde{A}$ and is evaluated by

$$c_r = \frac{\sum_{i=1}^{k-1} \underline{\mu}_{\tilde{A}}(x_i) \cdot x_i + \sum_{i=k+1}^{N} \bar{\mu}_{\tilde{A}}(x_i) \cdot x_i}{\sum_{i=1}^{k-1} \underline{\mu}_{\tilde{A}}(x_i) + \sum_{i=k+1}^{N} \bar{\mu}_{\tilde{A}}(x_i)} \tag{2.3}$$

using the well-known Karnik-Mendel algorithm [51], where $x \in \{x_1, x_2, \ldots, x_N\}$ and $x_{i+1} > x_i \ \forall i = 1 \text{ to } N-1$. Here $x = x_k$ is a switch point and $N$ denotes the number of sample points of $\bar{\mu}_{\tilde{A}}(x_i)$ and $\underline{\mu}_{\tilde{A}}(x_i)$.

**Definition 2.11** The centroid of an IT2FS is given by

$$c = \frac{(c_l + c_r)}{2} \tag{2.4}$$

where $c_l$ and $c_r$ are the left and the right end point centroids.

## 2.3 Proposed Approach

Given a time-series $c(t)$ for close price of a stock index, we observe consecutive 10 months' daily data for the above time-series to extract certain knowledge for prediction of the time series. To extract such knowledge, we partition the entire range of $c(t)$ into equal sized intervals $P_i$, $i = 1$ to $p$, and determine the list of possible changes in $c(t)$ from day $t = t_i$ to $t = t_{i+d}$ for any valid integer $i$ and a fixed delay $d$. Classical production rule-based reasoning [52] could be performed to predict the interval of $c(t_{i+d})$ from the known interval of $c(t_i)$ using the previously acquired rules. However, because of uncertainties in time-series, the strict production rules may not return the correct predictions. The logic of fuzzy sets, which has proved itself a successful tool to handle uncertainty, has therefore been used here to predict the membership of $c(t_{i+d})$ in a given partition $P_{i+d}$ from the measured membership of $c(t_i)$ in partition $P_i$.

In this chapter, each continuum neighborhood of data points of $c(t)$ in a given partition $P_i$ is represented by a T1 fuzzy set, and the union of all such T1 fuzzy sets under the partition is described by an IT2FS. The IT2FS model proposed for each individual partition can capture the uncertainty of the disjoint sets of data points within the partition. In addition, the transition: $c(t)$ to $c(t+d)$ from partition $P_i$ to partition $P_j$ is encoded as an IT2 fuzzy prediction rule, rather than a typical binary production rule. The IT2 prediction rule indicates that the linguistic variables present in the antecedent and consequent parts of the rule have IT2 MFs. The prediction of $c(t'+d)$ from a given measurement point $c(t')$, is done in two steps. In the first step, we use fuzzy reasoning to determine the membership of $c(t'+d)$ in partition $P'_j$ from the known membership of $c(t')$ in $P'_i$. After the inference is obtained, we use a T1/IT2 de-fuzzification depending on the type of reasoning used. The modality selection of reasoning (i.e., T1/IT2) is performed based on the distribution of data points in a given partition. This is undertaken in detail in the algorithm to be developed for rule identification from transitions history of data points in the time-series.

The principle of time-series forecasting introduced above is expected to offer good prediction accuracy, in case the time-series under consideration (called main factor) is not disturbed by external influences, such as changes in Government policies, macro/micro economic conditions, and many other unaccountable circumstances. Since all the external influences are not known, in many circumstances we model the influences by considering variation from other associated (secondary) world indices. Chen et al. [47] introduced an innovative approach to represent the effect of secondary indices to the main factor time-series. They considered composite variation of several secondary indices by measuring the deviation of individual index from the main factor time-series, and later used these deviations to determine normalized weights. These normalized weights are used later to scale the stock indices to determine the composite variation of secondary stock indices. To predict a stock data at day $t+1$ from the measurements of the same stock data at day $t$, Chen et al. determined the partition of the composite variation at day $t$ with an assumption that the main factor at day $t$ too would have the same partition. Later they used T1 fuzzy reasoning (using acquired rules in the training phase) to predict the stock data for the main factor at day $t+1$.

This chapter proposes three alternative approaches to economic time-series prediction. The first proposal considers employing IT2FS in place of T1 fuzzy reasoning introduced in [8–11, 21, 22, 47]. The IT2FS captures the inherent uncertainty in the time-series and thus provides a better fuzzy relational mapping from the measurement space to inference space, thereby offering better performance in prediction than its T1 counterpart. The second approach considers both IT2FS based reasoning along with feed-forward information from secondary stock indices, which usually are of more relative stability than the time-series under prediction. Thus the performance with feed-forward from secondary time-series gives better relative performance in comparison to the only IT2FS based reasoning. The third approach considers adaptation of T1 membership functions used to construct IT2FS MFs along with feed-forward connections from secondary stock indices. The performance of the third approach is better than its other two counterparts. The main steps of the algorithms are outlined below.

### 2.3.1 Training Phase

Given the MFTS and the SFTS of close price $c(t)$ for 10 months, we need to determine (i) group of type-2 fuzzy logical implications (prediction rules) for individual interval of main factor variation, and (ii) Secondary to Main Factor variation mapping. This is done by the following six steps.

1. Partitioning of main factor close price (MFCP) into p intervals (partitions) of equal length.
2. Construction of IT2 or T1 fuzzy sets as appropriate for each interval of close price.
3. IT2 or T1 fuzzy rule base selection for each interval.
4. Grouping of IT2/T1 fuzzy implication for individual main factor variation time-series $V_M^d(t)$.
5. Computing Composite Secondary Variation Series (CSVS) and its partitioning.
6. Determining secondary to main factor variation mapping.

Figures 2.1, 2.2, 2.3, 2.4, 2.5, 2.6 and 2.7 together explains the steps involved in the training phase. The details of individual steps are given below point-wise.

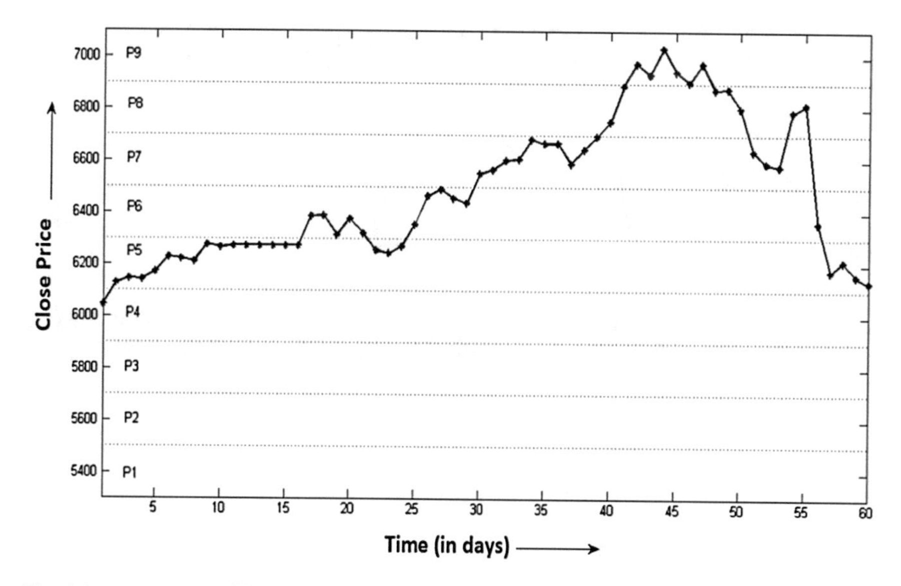


**Fig. 2.1** Time Series $c(t)$ and the partitions

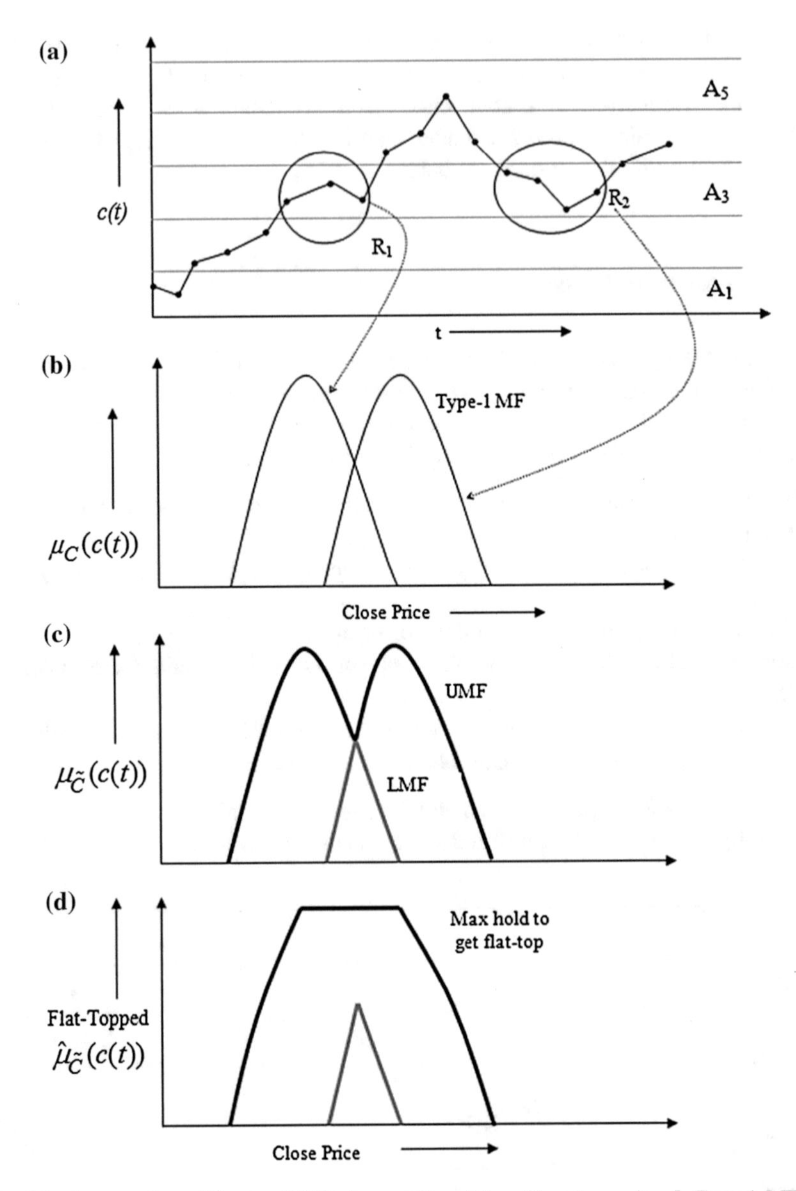


**Fig. 2.2** Construction of flat-top IT2FS for partition A3: **a** The close price, **b** Type-1 MFs for regions R1 and R2, **c** IT2FS representation of (**b**), **d** Flat-top IT2FS obtained by joining the peaks of two lobes

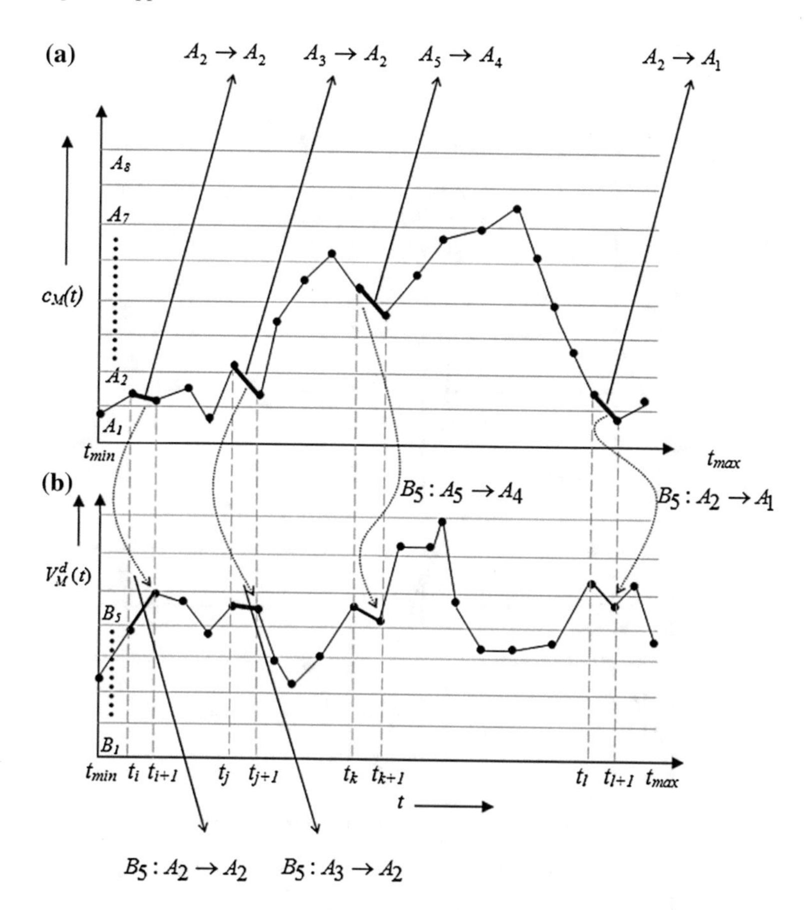


**Fig. 2.3** Construction of fuzzy logical implication and their grouping under MFVS $V_M^d(t)$ with d = 1: **a** If $c_M(t_i) \in A_k$ and $c_M(t_i + 1) \in A_j$; then the rule is $A_k \rightarrow A_j$, **b** If $V_M^d(t_i + 1) = B_s, \quad \exists s$, then grouping is done as $B_s{:}A_k \rightarrow A_j, \quad \exists j, k, s$

#### 2.3.1.1 Partitioning of Main Factor Close Prices into p Intervals of Equal Length

Consider a universe of discourse $U$ given by $[MIN - D_1,\ MAX + D_2]$, where MAX and MIN are the respective global maximum and global minimum of the time-series for close price $c(t)$ for a given duration of $t$ in [1, 10] months. $D_1$ and $D_2$ are positive real numbers in [1, 99], such that $(MAX + D_1)/100$ and $(MIN - D_2)/100$ are positive integers. Divide the universe $U$ into $p$ disjoint partitions: $P_1, P_2, \ldots, P_p$ of equal intervals as given in Fig. 2.1 [53] for more precision and clarity), where the length of an interval [47] is given by $[(MAX + D_1) - (MIN - D_2)]/p$.

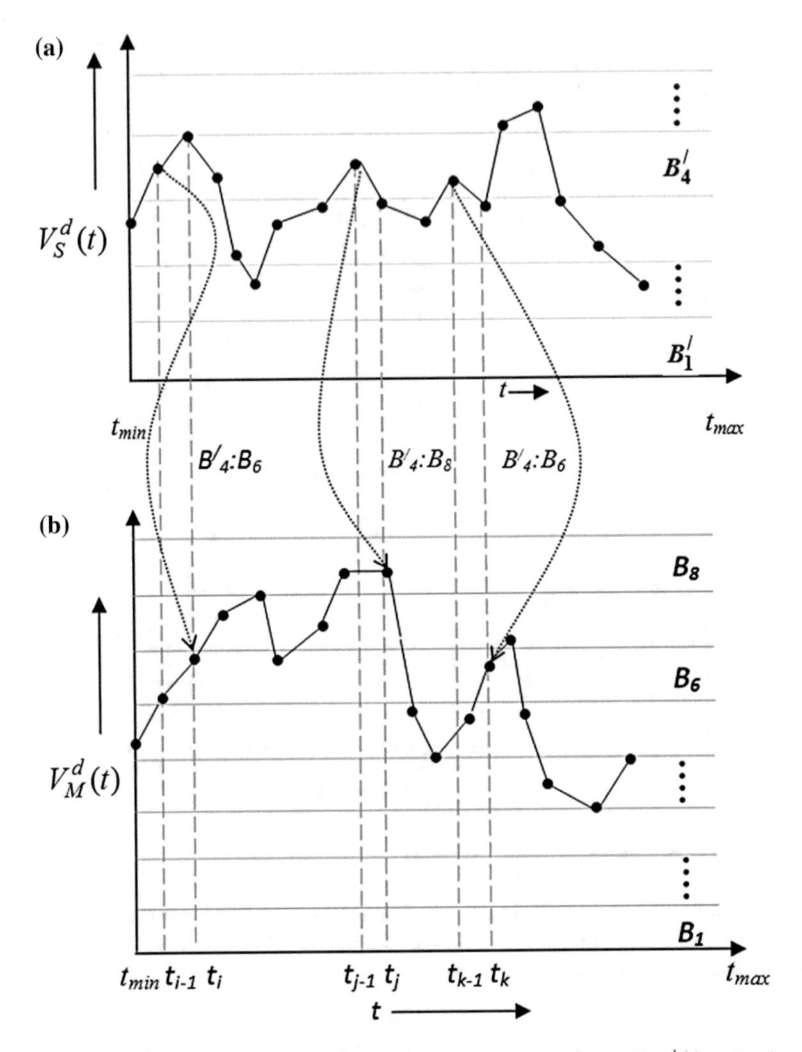


**Fig. 2.4** Secondary to Main factor variation mapping considering d = 1: If $V_M^d(t) \in B_k$, then the mapping is written as $B_j^{\prime} : B_k$

#### 2.3.1.2 Construction of IT2 or Type-1 Fuzzy Sets as Appropriate for Each Interval of Close Price

For each partition $P_i$, $i = 1$ to $p$ of $c(t)$, and for each set j of consecutive data points in $P_i$, we define fuzzy sets $A_{i,\,j}$ for $j = 1$ to $j_{Max}$, where the T1 MF of $A_{i,j}$ indicates the linguistic membership function (MF) CLOSE_TO_CENTRE_VALUE. For each group j of (three or more) consecutive data points of $c(t)$ in $P_i$, construct a

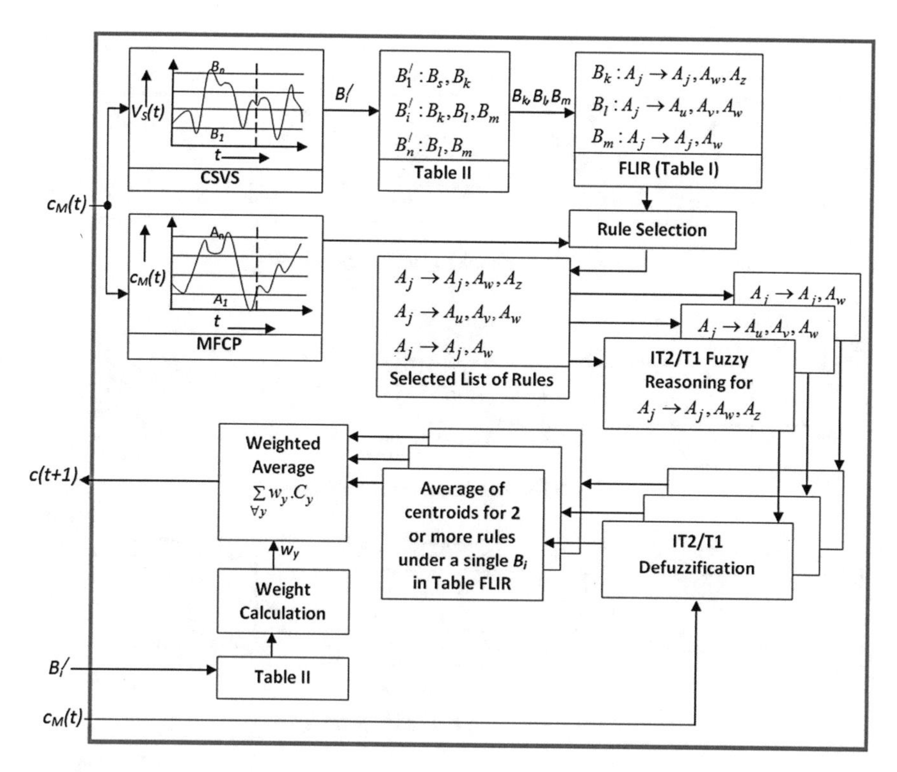


**Fig. 2.5** The main steps in the prediction algorithm of a stock index time-series considering d = 1

Gaussian T1 membership function with mean and standard deviation equal to the respective mean and standard deviation of these data points. Construct an IT2FS $\tilde{A}_i$, the footprint of uncertainty $\text{FOU}_i$ of which is obtained by taking the union of $A_{i,j}$ for $j = 1$ to $j_{Max}$. The constructed FOU is approximated (by joining the peaks of T1 MFs with a straight line of zero slope) with a flat top UMF to ensure convexity and normality [54, 55] of the IT2FS. The following special cases need to be handled for partitions with fewer data points.

If a partition $P_i$ includes only one data point of $c(t)$, we construct a T1 Gaussian MF with mean equal to the data point and very small variance of the order of $10^{-4}$ or smaller. If a partition $P_i$ includes only two consecutive data points of $c(t)$ we construct a T1 Gaussian MF with mean and standard deviation equal to the respective mean and standard deviation of these two data points. Lastly, if a partition $P_i$ includes only two (or more) discrete individual data points of $c(t)$ we construct two (or more) Gaussian MFs with means equal to the respective data points and very small variance of the order of $10^{-4}$ or smaller. We now construct a IT2FS by taking union of these T1 MFs.

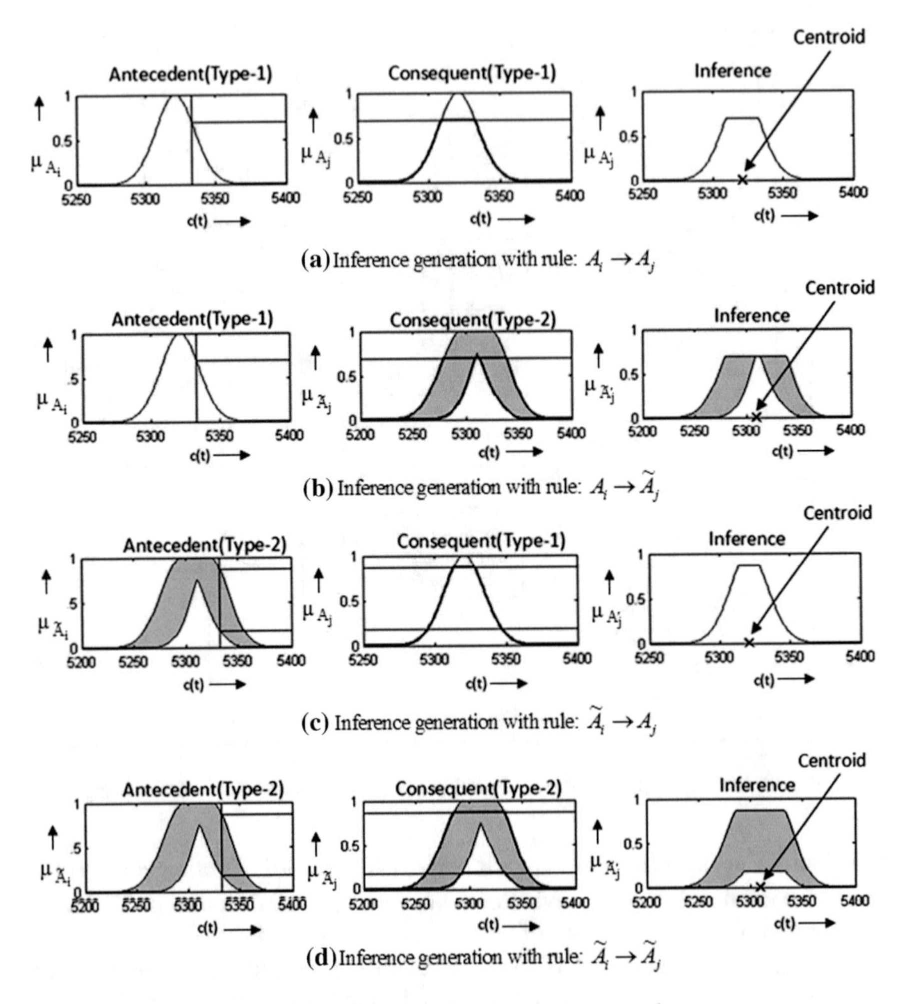


**Fig. 2.6** Inference Generation with T1/ IT2 antecedent-consequent pairs

#### 2.3.1.3 Fuzzy Prediction Rule (FPR) Construction for Consecutive $c(t)$s

For each pair of training days $t$ and $t + d$, we determine the mapping from $\tilde{A}_i$ to $\tilde{A}_j$, where $\tilde{A}_i$ and $\tilde{A}_j$ correspond to IT2 MF of the close prices at day $t$ and day $t + d$ respectively.

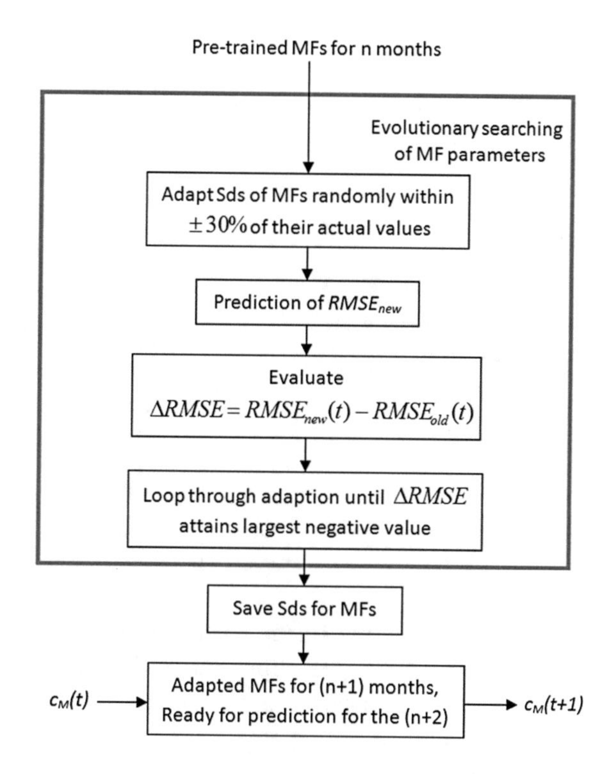


**Fig. 2.7** Optimal Selection of MFs to minimize RMSE

#### 2.3.1.4 Grouping of IT2/T1 Fuzzy Implications for Individual Main Factor Variation $V_M^d(t)$

(a) **MFVS Construction**: For trading days $t \in [t_{Min}, t_{Max}]$, we evaluate $V_M^d(t)$ using Eq. (2.1) for the main factor (here, TAIEX [56]).

(b) **Partitioning of $V_M^d(t)$ into $B_i$s**: Although in most of the practical cases $V_M^d(t)$ lies in $[-6\%, +6\%]$, we here consider a wider range of it in $[-\infty, +\infty]$, so as to not to disregard the possibility of occurrences of stray data points outside $[-6\%, +6\%]$. Partitioning the entire space of $[-\infty, +\infty]$ is performed by segregating the range: $[-6\%, +6\%]$ into equal sized intervals and the range beyond on either sides of it, i.e., $[-\infty, -6\%)$ and $(+6\%, +\infty]$ into two distinct intervals. Such partitioning ensures a uniformly high probability of occurrence of any data point at any one of the intervals for the band $[-6\%, +6\%]$ and a uniformly low probability of occurrence to any data point lying in $[(-\infty, -6\%)$ and $(+6\%, +\infty)]$ ranges. The entire space of $V_M^d(t)$ in $[-\infty, +\infty]$ is divided into 14 intervals (partitions) $B_1$ through $B_{14}$, where interval $B_1$ describes the range $[(-\infty, -6\%)$, $B_2$, through $B_{13}$ represent 12 partitions covering $[-6\%, +6\%]$ in order of increasing values of $V_M^d(t)$, and the interval $B_{14}$ represents the last partition $(+6\%, +\infty]]$.

**Table 2.1** Main factor fuzzy logical implication (FLI) considering d = 1

| Group | Time points | |
|---|---|---|
| | $t_1$ … $t_{i+1}$ | $t_{j+1}$ … $t_{k+1}$ … $t_{l+1}$ … $t_n$ |
| $B_1$ | … | … |
| … | … | … |
| $B_5$ | $A_2 \rightarrow A_2$ | $A_3 \rightarrow A_2A_5 \rightarrow A_4A_2 \rightarrow A_1$ |
| … | … | … |
| $B_{14}$ | … | … |

**Table 2.2** Main factor fuzzy logical implication (FLI) under variation groups considering d = 1

| Group | Antecedent of main factor | | | | |
|---|---|---|---|---|---|
| | $A_1$ | $A_2$ | $A_3$ … | $A_5$ … | $A_{18}$ |
| $B_1$ | … | … | … | … | |
| … | … | … | … | … | … |
| $B_5$ | | $A_2 \rightarrow A_2$<br>$A_2 \rightarrow A_1$ | $A_3 \rightarrow A_2$ | $A_5 \rightarrow A_4$ | |
| … | … | … | … | … | … |
| $B_{14}$ | … | … | … | … | … |

(c) **Grouping of FPRs Under Each Variation Group** $B_i$: For each feasible $t+d$ in $[t_{Min}, t_{Max}]$, find the partition $B_i$, such that $V_M^d(t+d)$ lies in the range of $B_i$. Also obtain the fuzzy sets $\tilde{A}_j, \tilde{A}_k$ corresponding to the partitions $P_j$ and $P_k$ at days $t$ and $t+d$ respectively. Then construct a rule: $\tilde{A}_j \rightarrow \tilde{A}_k$ with a label $B_i$, represented by

$$B_i : \tilde{A}_j \rightarrow \tilde{A}_k.$$

Repeat this step $\forall feasible\, t \in [t_{Min}, t_{Max}]$. Figure 2.3 describes the above mapping of $\tilde{A}_j \rightarrow \tilde{A}_k$ and its labeling against $B_i$ for a MFTS $c_M(t)$. Tables 2.1 and 2.2 clarifies the grouping of rules like $\tilde{A}_j \rightarrow \tilde{A}_k$ in $B_i$ following Fig. 2.3.

#### 2.3.1.5 Computing Composite Secondary Variation Series (CSVS) and Its Partitioning

(a) **Computing Secondary Factor Variation Series (SFVS)**: For *i*th elementary secondary factor $SF^i$, we evaluate $V_{S^i}^d(t)$ (variation in $SF^i$) using Eq. (2.1), where $c^i(t)$ and $c^i(t-d)$ denote the close price of *t*th day and $(t-d)$th day of *i*-th elementary secondary factor respectively.

(b) **Total Difference Variation** $(\Delta_{s^i})$ **Computation**: For i-th elementary secondary $SF^i$, evaluate total difference variation, denoted by $\Delta_{s^i}$ by using the following expression,

$$\Delta_{s^i} = \sum_{\forall t} |V_{S^i}^d(t-d) - V_M^d(t)| \tag{2.5}$$

where $V_M^d(t)$ and $V_{S^i}^d(t-d)$ denote the variation is main factor at day $t$ and that in $i$th SF[i] day $(t-d)$.

(c) **Normalization**: Use transformation (6) to obtain the normalized value of $\Delta_{s^i}$.

$$\hat{\Delta}_{s^i} = \left[\frac{\Delta_{s^i}}{\sum_{j=1}^{n} \Delta_{s^j}}\right]^{-1} = \frac{\sum_{j=1}^{n} \Delta_{s^j}}{\Delta_{s^i}} \tag{2.6}$$

where index $j$ in $[1, n]$ refer to different elementary SF[i]s.

(d) **Weight Computation**: Determine the normalized weighted variation for elementary SF over the training period (January 1 through October 31 of any calendar year).

$$W_S^i = \frac{\hat{\Delta}_{s^i}}{\sum_{j=1}^{n} \hat{\Delta}_{s^j}} \tag{2.7}$$

(e) **Composite Secondary Variation Series (CSVS) Computation** [47]: The overall variation at day t is given by

$$V_S^d(t) = \sum_{i=1}^{n} V_{s^i}^t \cdot W_{s^i}^t \tag{2.8}$$

#### 2.3.1.6 Determining Secondary to Main Factor Variation Mapping

Like the main factor time-series, the secondary factor variation series $(CSVS)$ is also partitioned into 14 intervals: $B'_1, B'_2, \ldots, B'_{14}$ following the same principle as introduced in step 4(b). For each $V_S^d(t-d)$ lying in $B'_i$, and for each $V_M^d(t)$ lying in $B_j, B_k, \ldots, B_l$, for all $t$, we group $B_j, B_k, \ldots, B_l$ under group $B'_i$.

Group $B'_i : B_j,\ B_k,\ \ldots, B_l$

Figure 2.4 illustrates the principle of group formation under $B'_i$. Here, for space limitation we show only 8 intervals $B_1$ through $B_8$ instead of 14 intervals. The frequency count of $B_j$ in MFTS at day $t$ for a given $B'_i$ in CSVS at day $t-d$ is evaluated in Fig. 2.4 and included in Table 2.3 for convenience.

**Table 2.3** Frequency of occurrence of main factor variation in each group of secondary factor variation considering d = 1

To Main Factor Variation →

From Secondary Factor Variation ↓

| | $B_1$ | $B_2$ | … | $B_6$… | | $B_8$ | … $B_{14}$ | |
|---|---|---|---|---|---|---|---|---|
| $B_1'$ | … | … | … | … | … | … | … | … |
| $B_2'$ | … | … | … | … | … | … | … | … |
| | … | … | … | … | … | … | … | … |
| $B_4'$ | 0 | 0 | … | 2 | … | 1 | … | 0 |
| | … | … | … | … | … | … | … | … |
| $B_{14}'$ | … | … | … | … | … | … | … | … |

## 2.3.2 *Prediction Phase*

Prediction of time-series at day $t+d$ from its close price at day $t$ could easily be evaluated by identifying all the rules having antecedent $A_j$, where $A_j$, denotes the fuzzy set corresponding to the partition at day $t$ in the partitioned main factor close price time-series. However, it is observed by previous researchers [8–11] that prediction using all the rules with $A_j$ as antecedent does not give good results. This chapter overcomes the above problem by selecting a subset of all possible rules with $A_j$ as antecedent. The subset-hood is determined by using the secondary variation time series. For example, if the partition returned by secondary factor time-series at day t is $B_i'$ then we obtain the corresponding partitions in MFTS by consulting the secondary to main factor variation series mapping introduced in Table 2.3. Suppose the Table 2.3 returns as the partitions in the MFTS. We now look for rules having antecedent $A_j$ in the labels in Table 2.2. The rules present under a given label are fired , and the average of the centroids of the resulting inferences is preserved. The weighed sum of the preserved average centroids (corresponding to individual labels) is declared as the predicted close price at day $t+d$. The algorithm for close price prediction is given below. Figure 2.5 provides the steps of the prediction algorithm schematically.

1. Obtain secondary variation $B_i'$ and main factor close price $A_j$ both for day $t$.
2. Using Table 2.3, determine $B_k$, $B_l$, $B_m$ etc. of main factor variation enlisted against $B_i'$.

3.

(a) **Rule Selection**: Identify fuzzy production rules with antecedent $A_j$ against rules with main factor variation $B_k, B_l, B_m$ in Table 2.2.

(b) For each production rule under a given $B_p$, $p \in \{k, l, m\}$.,

i. **IT2/T1 Fuzzy Reasoning**: Perform IT2/T1 fuzzy reasoning with rules $A_j \to A_u$, $A_j \to A_v$, $A_j \to A_w$ if the row $B_p$ in Table 2.2 includes the rule $A_j \to A_u, A_v, A_w$. If the group $B_p$ does not contain any rule then we consider the mid value of the partition corresponding to the partition $A_j$ for forecasting following [47].

ii. **IT2/T1 Defuzzification:** Employ IT2 or T1 defuzzification, as applicable, to obtain centroids of the discretized MFs: $A_u$, $A_v$, $A_w$ and take the average of the centroids.

The procedure of reasoning and defuzzification considering the presence of T1/IT2 MFs in antecedent/consequent is given separately for convenience of the readers.

(c) **Weight Calculation and Prediction**: Determine the frequency counts $f_k, f_l, f_m$ of the main factor variation $B_k$, $B_l$, $B_m$ under secondary variation $B_i'$ in Table 2.3 to determine the probability of occurrences as $f_k/(f_k+f_l+f_m)$, $f_l/(f_k+f_l+f_m)$, $f_m/(f_k+f_l+f_m)$. We use the probabilities to defuzzify expected value of main factor close price for the next day by taking sum of products of probabilities and average centroids values under $B_k, B_l$, $B_m$.

The complete steps of prediction of a time series are illustrated in Fig. 2.5.

**Procedure of T1/IT2 Reasoning and Defuzzification**

***Case I: When Both Antecedent and Consequents are IT2FS***

(a) **IT2 Reasoning**: Let $\tilde{A}_i(c(t))$ be the value of $\tilde{A}_i$ for linguistic variable $x = c(t)$. Let UMF and LMF for $\tilde{A}_i$ be $UMF_i(c(t))$ and $LMF_i(c(t))$ respectively. On firing the rules: $\tilde{A}_i \to \tilde{A}_j$, $\tilde{A}_i \to \tilde{A}_k$ and $\tilde{A}_i \to \tilde{A}_l$ we determine the fuzzy inferences by the following procedure. For the rule: $\tilde{A}_i \to \tilde{A}_j$, the IT inference is obtained by

$$UMF_j' = Min\,[UMF_i(c'),\, UMF_j], \tag{2.9}$$

$$LMF_j' = Min\,[LMF_i(c'),\, LMF_j], \tag{2.10}$$

where $c'$ is a measured value of $c(t)\,\exists t$.

Similarly, we obtain $UMF_y'$ and $LMF_y'$ by replacing index j by $y$ for $y \in \{k, l\}$ for the remaining rules.

(b) **IT2 Centroid Computation**: For each pair of $UMF_x'$ and $LMF_x'$ for $x \in \{j, k, l\}$, we determine the centroid of IT2FS $A_x'$ by the following method. Determine the lower End Point centroid $c_L^x$ and Upper End Point centroid $c_R^x$, for $x \in \{j, k, l\}$ and centroid $c^x$ following Eqs. 2.2, 2.3 and 2.4 respectively. Thus for the rule: $B_i : \tilde{A}_i \to \tilde{A}_j, \tilde{A}_k, \tilde{A}_l$, we obtain the $c^j$, $c^k$ and $c^l$. Determine the average of $c^j$, $c^k$ and $c^l$.

***Case II: When Antecedent is T1FS and Consequent is IT2FS***

(a) **IT2 Reasoning**: For the rule: $B_i : A_i \to \tilde{A}_j, \tilde{A}_k, \tilde{A}_l$ we determine the fuzzy inferences by the following procedure for

$$UMF_j' = Min\left[\mu_{A_i}(c'),\ UMF_j\right], \tag{2.11}$$

$$LMF_j' = Min\left[\mu_{A_i}(c'),\ LMF_j\right], \tag{2.12}$$

where $c'$ is a measured value of $c(t)\, \exists t$ and $\mu_{A_i}(c')$ is the T1 membership at $c(t) = c'$. Similarly, we obtain $UMF_y'$ and $LMF_y'$ by replacing index $j$ by $y$ for $y \in \{k, l\}$.

(b) **IT2 Centroid Computation**: The centroid computation procedure, here, is similar to that in Case I.

***Case III: When Antecedent is IT2FS and Consequent is T1 FS***

(a) **T1 Reasoning**: For the rule: $B_i : \tilde{A}_i \to A_j, A_k, A_l$, we determine the fuzzy inferences by the following procedure.

$$\mu_{A_j'}(c(t)) = Min\left[UMF_i(c'), \mu_{A_j}(c(t))\right], \tag{2.13}$$

where $c'$ is a measured value of $c(t)\, \exists t$.

Similarly, we obtain $\mu_{A_y'}(c(t))$ by replacing index $j$ by $y$ for $y \in \{k, l\}$.

(b) **T1 Centroid Computation**: For each T1 discretized MF $\mu_{A_x'}(c(t))$ for $x \in \{j, k, l\}$, we determine the T1 centroid of $A_x'$ by the following formula

$$c^x = \frac{\sum_{c(t_i)=-\infty}^{\infty} \mu_{A_x}(c(t_i)).c(t_i)}{\sum_{c(t_i)=-\infty}^{\infty} \mu_{A_x}(c(t_i))} \quad \text{For } x \in \{j, k, l\} \tag{2.14}$$

Thus for the rule $B_i : \tilde{A}_i \to A_j, A_k, A_l$, we obtain the centroids $c^j$, $c^k$ and $c^l$. Determine the average of $c^j$, $c^k$ and $c^l$.

***Case IV: Antecedent and Consequent Both are T1 FS***

(a) **T1 Reasoning**: For the rule: $B_i : A_i \rightarrow A_j, A_k, A_l$ we determine the fuzzy inferences by the following procedure.

$$\mu_{A'_j}(c(t)) = Min\,[\mu_{A_i}(c'), \mu_{A_j}(c(t))],$$

Where $c'$ is a measured value of $c(t)\,\exists t$.
Similarly, we obtain $\mu_{A'_y}(c(t))$ by replacing index j by y for $y \in \{k, l\}$.

(b) **T1 Centroid Computation**: The centroid computation procedure, here, is similar to that in Case III.

Figure 2.6 provides the inference generation mechanism introduced above graphically for the above four cases.

### 2.3.3 Prediction with Self-adaptive IT2/T1 MFs

Large scale experiments with time-series prediction reveal that the results in RMSE are highly influenced by the shape of MFs used in the antecedent/consequent of the prediction rules. This motivated us to arrange on-line selection of MFs from a standard list, here triangular and Gaussian, with provisions for variations in their base-width. The optimal selection of base width can be performed by employing an evolutionary algorithm with an ultimate aim to minimize the RMSE. Any standard evolutionary/swarm algorithm could serve the above purpose. However, for our experience of working with Differential Evolution (DE) algorithm [57, 58] coupled with its inherent merits of low computational overhead, simplicity, requirement of fewer control parameters and above all its high accuracy, we used DE to adaptively select the right structure of MFs with RMSE as the fitness function.

Figure 2.7 provides a schematic overview of the MF adaptation scheme. The bold box in Fig. 2.7 includes the complete adaptation steps, while the bottommost block represents the prediction algorithm with adapted parameters. The adaptation module makes trial selection of standard deviation (base-width) of the Gaussian (triangular) MFs within ±30% of their original values. Next the change in RMSE due to adoption of the new MFs is evaluated. Finally, we loop through the above steps until no further reduction in change in RMSE is observed. The last obtained values of parameters of MFs are saved for subsequent usage in prediction. The benefits of the adaptation of MFs is compared in the next section vide Fig. 2.8.

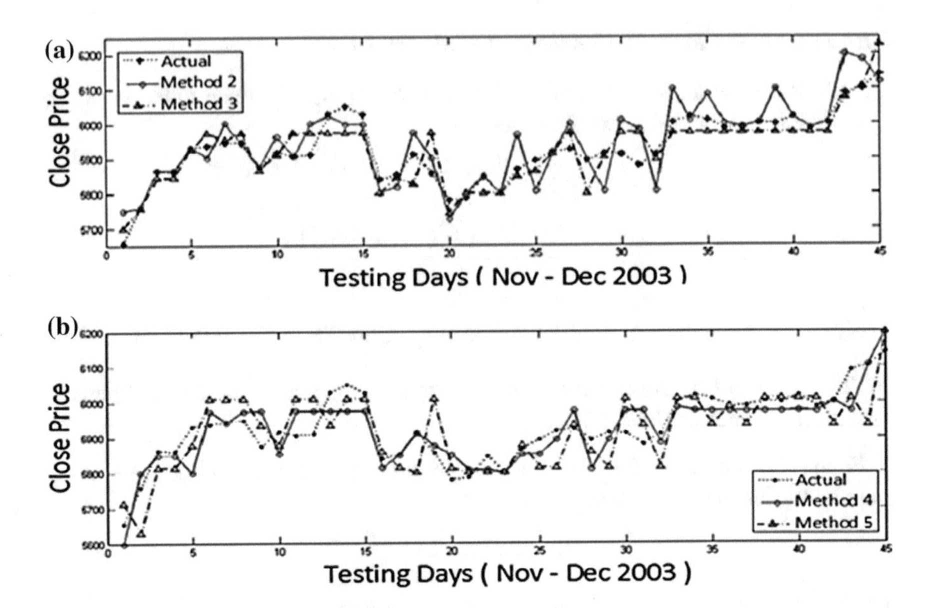


**Fig. 2.8** Forecasted TAIEX of the months November and December 2003 using Gaussian MFs and Triangular MFs, **a** Actual TAIEX, forecasted data using proposed method 2 without adopting the Gaussian MFs and forecasted data using proposed method 3 adopting the Gaussian MFs, **b** Actual TAIEX, forecasted data using proposed method 4 without adopting the triangular MFs and forecasted data using proposed method 5

## 2.4 Experiments

The experiment includes both training and testing with TAIEX [56] close price [13], hereafter called main factor, and NASDAQ and DOWJONES close prices as secondary factors. The training session comprises T1/IT2 membership function construction, extraction of fuzzy prediction rules and mapping of secondary to main factor variations. The testing session comprises fuzzy prediction rule selection for firing, T1/IT2 fuzzy reasoning as applicable, defuzzification and weighted averaging of multiple defuzzified rules falling under different main factor variations. While performing the experiments, we consider five distinctive methods. The proposed method 1 includes only IT2FS based reasoning, ignoring the effect of secondary factor without adaptation of membership function is given only for the sake of academic interest. The proposed method 2 to proposed method 5 includes the influence of secondary factor. The proposed method 2 to proposed method 5 are hereafter called (a) Proposed method 2: fixed Gaussian MF (without adaptation), (b) Proposed method 3: Gaussian with provisions for adaptation in standard deviation, (c) Proposed method 4: fixed triangular MF (without adaptation), and (d) Proposed method 5: triangular with provisions for adaptation in base-width. The training and prediction algorithms incorporating the above five types of MFs are hereafter referred to as proposed methods: 1, 2, 3, 4 and 5 respectively for brevity.

**Table 2.4** Strategies adopted in various experimental proposed methods

| Methods | Reasoning | Membership function considered | Secondary factor considered | Adaptation |
|---|---|---|---|---|
| Proposed method 1 | IT2 | Gaussian | No | No |
| Proposed method 2 | T1 and IT2 combined | Gaussian | Yes | No |
| Proposed method 3 | T1 and IT2 combined | Gaussian | Yes | Yes |
| Proposed method 4 | T1 and IT2 combined | Triangular | Yes | No |
| Proposed method 5 | T1 and IT2 combined | Triangular | Yes | Yes |

For clarity, we summarized the strategies adopted in these five methods are shown in Table 2.4. The initial MFs in Proposed method 4 and Proposed method 5 are represented by isosceles triangles with a peak membership of one at the centre and base-width equal to $6\sigma$, where $\sigma$ denotes the standard deviation of the consecutive data points in a given partition of close price.

## 2.4.1 *Experimental Platform*

The experiment was performed using MATLAB 2012b under WINDOWS-7 operating system running on a IBM personal computer with Core i5 processor with system clock of 3.60 G-Hz frequency and system RAM of 8 GB.

## 2.4.2 *Experimental Modality and Results*

### 2.4.2.1 Policies Adopted

The close price data for both main factor and secondary factors are obtained for the period 1990–2004 from the website [56]. The training session was fixed for 10 months: January 1 to October 31 of each year on all trading days. In case all the trading days of secondary factors do not coincide with those of the main factor, we adopt two policies for the following two cases. Let Set A and B denote the dates of trading in main and secondary factors respectively. If $A - B$ (A minus B) is a non-null set, then the close price of previous trading days in secondary factor has to be retained over the missing (subsequent) days. If $B - A$ is non-null set, then we adopt the following policies. First, if the main factor has missing trading days due to holidays and/or other local factors, then no training is undertaken on those days.

Second, in the trading of next day of main factor, we consider the influence of the last day of trading in secondary closing price. After the training is over, the following items including prediction rules (also called Fuzzy Logic Implications (FLI)) and secondary to main factor variation groups are saved for the subsequent prediction phase. The prediction was done for each trading day during the month of November and December. Comparison of the results of prediction with those of Chen et al. [47] is given in authors' webpage [53], and is not given here for space restriction. The results of prediction (November-December, 2003) with and without adaptation of parameters (standard deviations) of MFs are given in Fig. 2.8 along with the actual close price.

#### 2.4.2.2 MF Selection

Experiments are performed with both Gaussian and triangular T1 MFs. The UMF (LMF) of the IT2FS is obtained by taking maximum (Minimum) of the T1 MFs describing the same linguistic concept obtained from different sources. Figure 2.2 respectively provides the construction of IT2FS from triangular and Gaussian T1 MFs, following the steps outlined in Sect. 2.3. The relative performance of triangular and Gaussian MFs is examined by evaluating RMSE of the predicted close price with respect to its actual TAIEX values. In most of the test cases, prediction of close price is undertaken during the months of November and December of any calendar year between 1999 and 2004.

The RMSE plots shown in Fig. 2.8 reveal that triangular MFs yield better prediction results (less RMSE) than its Gaussian counterpart. For example, the RMSE for TAIEX for the year 2003 using triangular and Gaussian MFs are respectively found to be 37.123 and 47.1108 respectively, justifying the importance of triangular MFs over Gaussian ones in the time-series prediction.

#### 2.4.2.3 Adaptation Cycle

The training algorithm is run with the close price time-series data from January 1st to October 31st on all trading days. For tuning the T1 MFs (before IT2FS construction) for qualitative prediction, the adaption algorithm is run for the period of September 1st to October 31st for the subsequent prediction of November. After the prediction of November month is over, the adaption procedure is again repeated for the month of October 1st to November 30th in order to predict the TAIEX close price in December. Such adaption over two consecutive months is required to track any abnormal changes (such as excessive level shift) in the time-series.

The improvement in performance due to inclusion of adaptation cycles is introduced in Fig. 2.8 (see [53] for precision), obtained by considering Gaussian MFs. It is apparent from Fig. 2.8a that in presence of adaptation cycles, the RMSE appears to be 47.1108, while in absence of adaptation, RMSE is found to be 52.771. The changes in results (RMSE) in presence of adaptation cycles due to use of

**Table 2.5** Comparison of RMSEs obtained by the proposed technique with existing techniques

| Methods | Years | | | | | | |
|---|---|---|---|---|---|---|---|
| | 1999 | 2000 | 2001 | 2002 | 2003 | 2004 | Mean |
| 1. Huarng et al. [36] (Using NASDAQ) | NA | 158.7 | 136.49 | 95.15 | 65.51 | 73.57 | 105.88 |
| 2. Huarng et al. [36] (Using Dow Jones) | NA | 165.8 | 138.25 | 93.73 | 72.95 | 73.49 | 108.84 |
| 3. Huarng et al. [36] (Using $M_{1b}$) | NA | 160.19 | 133.26 | 97.1 | 75.23 | 82.01 | 111.36 |
| 4. Huarng et al. [36] (Using NASDAQ and $M_{1b}$) | NA | 157.64 | 131.98 | 93.48 | 65.51 | 73.49 | 104.42 |
| 5. Huarng et al. [36] (Using Dow Jones and $M_{1b}$) | NA | 155.51 | 128.44 | 97.15 | 70.76 | 73.48 | 105.07 |
| 6. Huarng et al. [36] (Using NASDAQ, Dow Jones and $M_{1b}$) | NA | 154.42 | 124.02 | 95.73 | 70.76 | 72.35 | 103.46 |
| 7. Chen et al. [21, 31, 32] | 120 | 176 | 148 | 101 | 74 | 84 | 117.4 |
| 8. U_R model [31, 32] | 164 | 420 | 1070 | 116 | 329 | 146 | 374.2 |
| 9. U_NN model [31, 32] | 107 | 309 | 259 | 78 | 57 | 60 | 145.0 |
| 10. U_NN_FTS model [27, 31, 32] | 109 | 255 | 130 | 84 | 56 | 116 | 125.0 |
| 11. U_NN_FTS_S model [27, 31, 32] | 109 | 152 | 130 | 84 | 56 | 116 | 107.8 |
| 12. B_R model [31, 32] | 103 | 154 | 120 | 77 | 54 | 85 | 98.8 |
| 13. B_NN model [31, 32] | 112 | 274 | 131 | 69 | 52 | 61 | 116.4 |
| 14. B_NN_FTS model [31, 32] | 108 | 259 | 133 | 85 | 58 | 67 | 118.3 |
| 15. B_NN_FTS_S model [31, 32] | 112 | 131 | 130 | 80 | 58 | 67 | 96.4 |
| 16. Chen et al. [47] (Using Dow Jones) | 115.47 | 127.51 | 121.98 | 74.65 | 66.02 | 58.89 | 94.09 |
| 17. Chen et al. [47] (Using NASDAQ) | 119.32 | 129.87 | 123.12 | 71.01 | 65.14 | 61.94 | 95.07 |
| 18. Chen et al. [47] (Using $M_{1b}$) | 120.01 | 129.87 | 117.61 | 85.85 | 63.1 | 67.29 | 97.29 |
| 19. Chen et al.[47] (Using NASDAQ and Dow Jones) | *116.64* | *123.62* | *123.85* | *71.98* | *58.06* | *57.73* | *91.98* |
| 20. Chen et al. [47] (Using Dow Jones and $M_{1b}$) | 116.59 | 127.71 | 115.33 | 77.96 | 60.32 | 65.86 | 93.96 |
| 21. Chen et al. [47] (Using NASDAQ and $M_{1b}$) | 114.87 | 128.37 | 123.15 | 74.05 | 67.83 | 65.09 | 95.56 |
| 22. Chen et al. [47] (Using NASDAQ, Dow Jones and $M_{1b}$) | 112.47 | 131.04 | 117.86 | 77.38 | 60.65 | 65.09 | 94.08 |
| 23. Karnik-Mendel [44] induced stock prediction | 116.60 | 128.46 | 120.62 | 78.60 | 66.80 | 68.48 | 96.59 |

(continued)

**Table 2.5** (continued)

| Methods | Years | | | | | | |
|---|---|---|---|---|---|---|---|
| | 1999 | 2000 | 2001 | 2002 | 2003 | 2004 | Mean |
| 24. Chen et al. [50] (Using NASDAQ, Dow Jones and $M_{1b}$) | 101.47 | 122.88 | 114.47 | 67.17 | 52.49 | 52.84 | 85.22 |
| 25. Chen et al. [59] | 87.67 | 125.34 | 114.57 | 76.86 | 54.29 | 58.17 | 86.14 |
| 26. Cai et al. [60] | 102.22 | 131.53 | 112.59 | 60.33 | 51.54 | 50.33 | 84.75 |
| 27. Mu-Yen Chen [61] | NA | 108 | 88 | 60 | 42 | NA | 74.5 |
| 28. Proposed method 1 | 114.20 | 127.12 | 110.50 | 70.56 | 52.10 | 48.40 | 87.18 |
| 29. Proposed method 2 | 101.84 | 125.87 | 111.60 | 71.66 | 52.77 | 50.166 | 85.654 |
| 30. Proposed method 3 | 92.665 | 108.18 | 105.51 | 66.99 | 47.11 | 43.83 | 77.382 |
| 31. Proposed method 4 | 94.610 | 113.04 | 110.81 | 66.04 | 48.77 | 46.179 | 79.910 |
| 32. Proposed method 5 | 89.021 | 99.765 | 101.71 | 58.32 | 37.12 | 36.600 | 70.424 |

triangular MFs are illustrated in Fig. 2.8b. Both the realizations confirm that adaptation has merit in the context of prediction, irrespective of the choice of MFs.

#### 2.4.2.4 Varying *d*

We also study the effect of variation of '*d*' [62, 63] on the results of forecasting using proposed method 5. Here, for each integer value of d in [1, 4], we obtain the plots of actual and forecasted close price as indicated in Fig. 2.5. The fuzzy logical implication rules and frequency of occurrences from CSVS to MFVS are determined using Fig. 2.9 following similar approach as done for $d = 1$. These rules and frequency of occurrences are given in Tables 2.7 and 2.14 (SEE APPENDIX). It is apparent from Fig. 2.9 that forecasted price with delay $d = 1$ yields an RMSE of 36.6006, which is found to be smallest among the considered RMSEs for d = 1, 2, 3 and 4. This indicates that setting $d = 1$ returns the best possible prediction, which also has logical justification in the sense that the predicted close price intricately depends on close price of yesterday, rather than that of day before yesterday or its preceding days.

## 2.5 Performance Analysis

This section attempts to compare the relative performance of the proposed five techniques with 27 other techniques [21, 27, 31, 32, 36, 47] using RMSE as the metric for comparison. Table 2.5 provides the results of comparison for the period 1999–2004 with mean and standard deviation of all the RMSEs obtained for the above period. It is apparent from Table 2.5 that the entries in the last row are

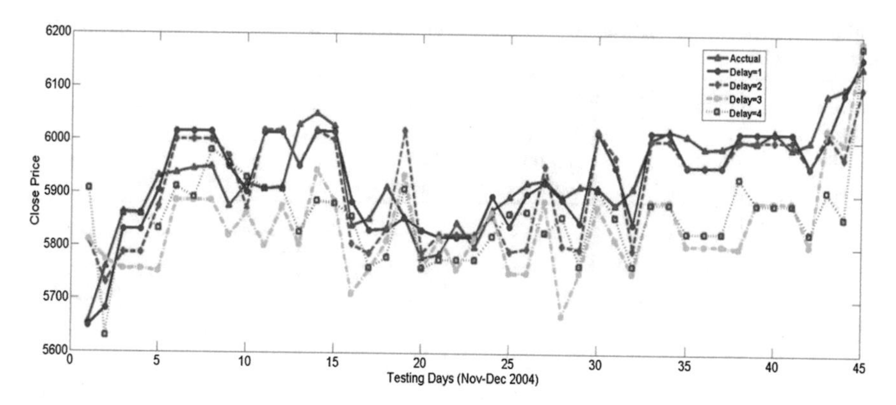


**Fig. 2.9** Forecasted TAIEX of the months November and December 2004 using Proposed method 5 for different values of d. The respective RMSEs are RMSE = 36.6006 for d = 1, RMSE = 72.8012 for d = 2, RMSE = 122.4201 for d = 3, RMSE = 140.2005 for d = 4

smaller than the entries above. This indicates that that RMSE for each column on the last row of Table 2.5 being the smallest, the proposed method 5 seems to outperform the other techniques (calculated with respect of mean of 6 years RMSE) by at least 23%, encountered in method-19 in Table 2.5.

We here use paired t-test [64] to examine the statistical confidence on the results of prediction by different algorithms using RMSE as the metric. Let, $H_o$ be the null hypothesis to compare two algorithms' performance, where one is the reference algorithm, while the other is any one of the existing algorithms. Here, we consider the proposed algorithm as the reference algorithm. Thus, $H_o =$ Performance of algorithm $A$ and reference algorithm $R$ are comparable.

Let A be the algorithm by Chen et al. [47]. To statistically validate the Hypothesis $H_o$, we evaluate t-measure, given by

$$t = \frac{(m_A - m_R)}{\sqrt{{s_A}^2 + {s_R}^2}}, \tag{2.15}$$

where $m_A$ and $m_R$ are the mean values of the distributions of RMSE obtained by algorithms A and R respectively with equal sample size in the two distributions, and $s_A$ and $s_R$ are the standard deviations of the respective samples obtained by algorithms $A$ and $R$.

After evaluation of statistic t, we consult the t-Table (Table 2.6) with degrees of freedom $KI =$ sample size of any one population minus $1 = n - 1$, say. Let the value obtained from the t-Table for given confidence level $\alpha$ and $KI$ be $z$. Now, if $z < t$, the calculated value by formula (2.15), then the $H_o$ is wrong, and its contradiction that the proposed algorithm is better than A with respect to RMSE is true. We now repeat the above steps for different comparative algorithms A and found that $z < t$ always holds, thereby indicating that the proposed algorithm outperforms all other existing algorithms.

**Table 2.6** Results of statistical significance with the proposed method 1–5 as the reference, one at a time (t-table)

| | Statistical significance | | | | |
|---|---|---|---|---|---|
| Existing methods | | Reference methods | | | |
| | Proposed method 1 | Proposed method 2 | Proposed method 3 | Proposed method 4 | Proposed method 5 |
| 1 | – | – | + | + | + |
| 2 | + | + | + | + | + |
| 3 | + | + | + | + | + |
| 4 | + | + | + | + | + |
| 5 | + | + | + | + | + |
| 6 | – | + | + | + | – |
| 7 | – | + | + | + | + |

(continued)

**Table 2.6** (continued)

| Existing methods | Statistical significance | | | | |
|---|---|---|---|---|---|
| | | Reference methods | | | |
| | Proposed method 1 | Proposed method 2 | Proposed method 3 | Proposed method 4 | Proposed method 5 |
| 8 | + | + | − | + | − |
| 9 | + | − | − | − | − |
| 10 | − | − | + | − | − |
| 11 | − | + | + | + | + |
| 12 | + | + | + | + | + |
| 13 | + | − | − | − | − |
| 14 | − | − | − | − | − |
| 15 | + | + | + | + | + |
| 16 | + | + | + | + | + |
| 17 | + | + | + | − | + |
| 18 | + | + | + | + | + |
| 19 | + | − | − | + | + |
| 20 | + | + | − | + | + |
| 21 | + | + | − | + | + |
| 22 | − | − | + | − | + |
| 23 | + | + | + | + | + |
| 24 | − | + | + | + | − |
| 25 | − | + | + | + | + |
| 26 | − | + | − | + | + |
| 27 | + | + | + | − | + |

Table 2.6 is designed to report the results of statistical test considering proposed method 1–5 as the reference. The degree of freedom is here set to 5 as the prediction data set used involves six years' RMSE data. The plus (minus) sign in Table 2.6 represents that the difference of means of an individual method with the proposed method as reference is significant (not significant). The degree of significance here is studied at 0.05 level, representing 95% confidence level.

## 2.6 Conclusion

This chapter introduced a novel approach to stock index time-series prediction using IT2Fs. Such representation helps overcoming the possible hindrances in stock index prediction as introduced in the introduction. Both triangular and Gaussian MFs along with provision of their adaptation have been introduced to examine their relative performance in prediction. The strategy used to consider secondary to main factor variation has considerably improved the relative performance of the stock index time-series prediction. A thorough analysis of results using RMSE as the metric indicates that the proposed methods outperform the existing techniques on stock index prediction by a considerable margin ($\geq$23%). Out of the five proposed methods, the method employing triangular MF with provision for its adaptation yields the best performance following the prediction of TAIEX stock data for the period of 1999–2004 with DOWJONES and NASDAQ together as the composite secondary index. A statistical analysis undertaken with paired t-test confirms that each of the proposed algorithms outperforms most of the existing algorithms with root mean square error as the key metric at 95% confidence level. With an additional storage of fuzzy logical implication rules and frequency of occurrences from CSVS to MFVS for $d = 1, 2, \ldots, k$, we would be able to predict the close price on the next day, next to next day and the like from today's close price. Further extension of the proposed technique can be accomplished by using General Type-2 fuzzy sets, which is expected to improve performance at the expense of additional complexity.

## 2.7 Exercises

1. Graphically plot the interval type-2 fuzzy set constructed from type-2 membership functions in Fig. 2.10.

   [**Hints:** The UMF and LMF constructed from the given type-1 MFs are given in Fig. 2.11.]

2. Construct the rules from a partitioned time-series, indicated in Fig. 2.12.

   [**Hints:** The rules following the occurrence of the data point in the partition are: $P_1 \rightarrow P_2$, $P_2 \rightarrow P_4$, $P_4 \rightarrow P_2$, $P_2 \rightarrow P_1$.]

3. Let there be three partitions $P_1$, $P_2$, $P_3$ of a stock data of a stock data, the corresponding fuzzy sets are $A_1$, $A_2$, and $A_3$. Suppose we have the rules: $A_1 \rightarrow A_2$ and $A_1 \rightarrow A_3$ as indicated below, Determine the stock price of tomorrow if the stock price of today, as indicated falls in partition $P_1$ (i.e. fuzzy set of $P_1$). Presume that, $\sum_x x\_movement$ of the inferred membership function is 100 with

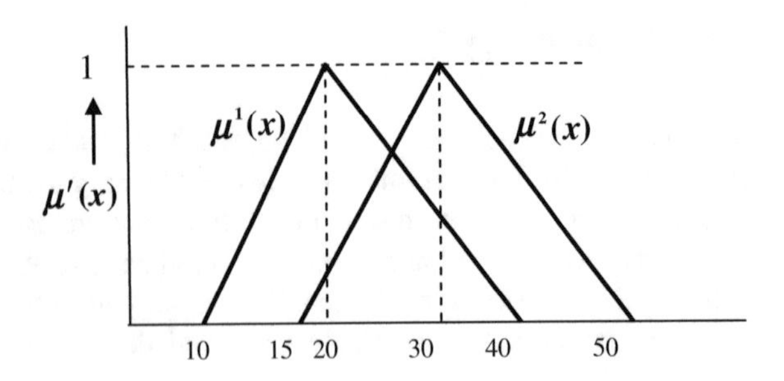


Fig. 2.10 Figure for Problem 1

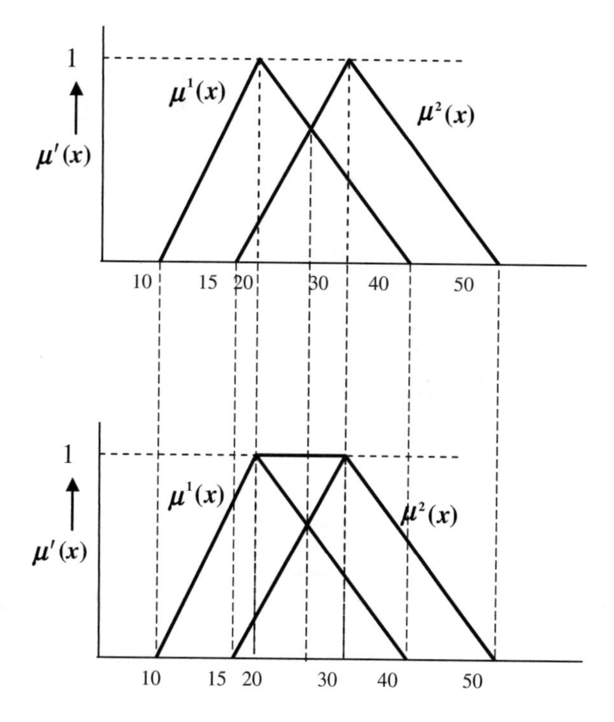


**Fig. 2.11** Solution for Problem 1

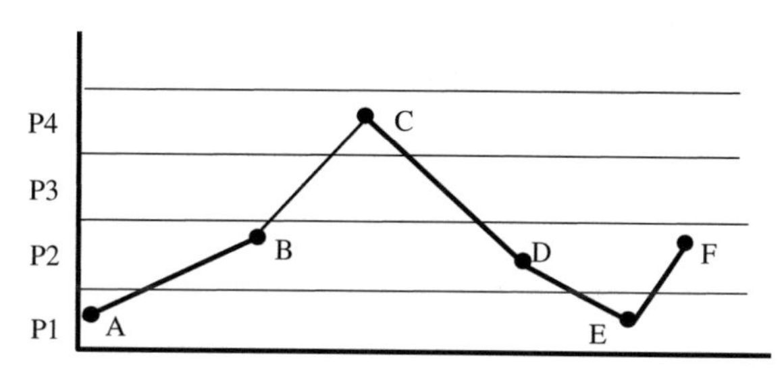


**Fig. 2.12** Figure for Problem 2

and area under the inferred membership = 12 unit [Ans: 100/12 = 8.33] (Fig. 2.13).

4. Let the inferred membership function be a sine function for x = 0 to π. Find the centroid. Refer Figs. 2.14 and 2.15.

$$[\textbf{Hints}: \text{Centroid} = \frac{\int_0^{\pi} x \sin x dx}{\int_0^{\pi} \sin x dx} = \frac{[x(-\cos x) + \sin x]_0^{\pi}}{[-\cos \pi]_0^{\pi}} = \frac{\pi}{1+1} = \pi/2]$$

5. Let the partitions be $P_1, P_2$ and, also the IT2FS used for three partitions be $\tilde{A}_1, \tilde{A}_2$ and $\tilde{A}_3$ respectively. Given the IT2FS for the stock data and the rules $A_1 \to A_2$ and $A_1 \to A_3$. If today's stock price falls in $\tilde{A}_3$, then will you be able to generate the fuzzy inference for tomorrow.

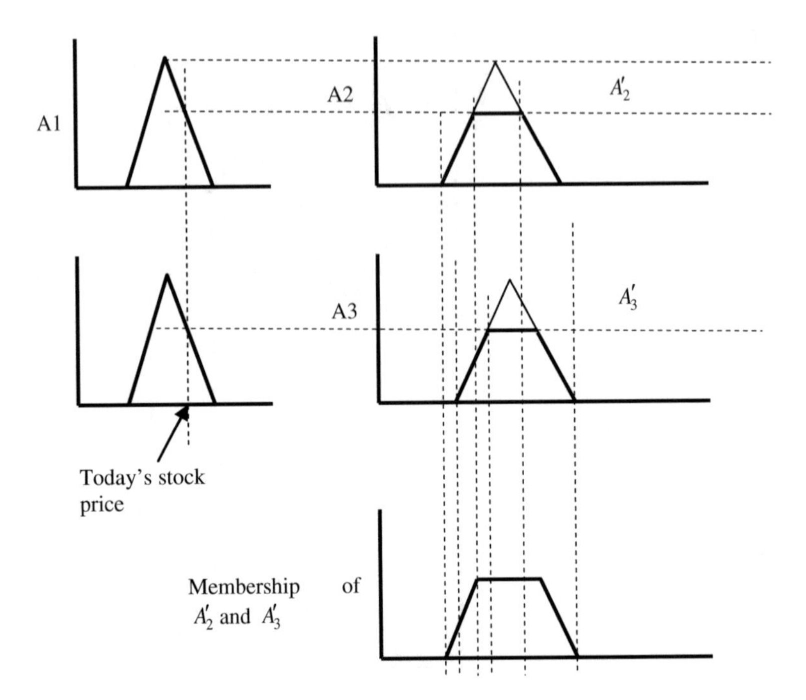


**Fig. 2.13** Figure for Problem 3

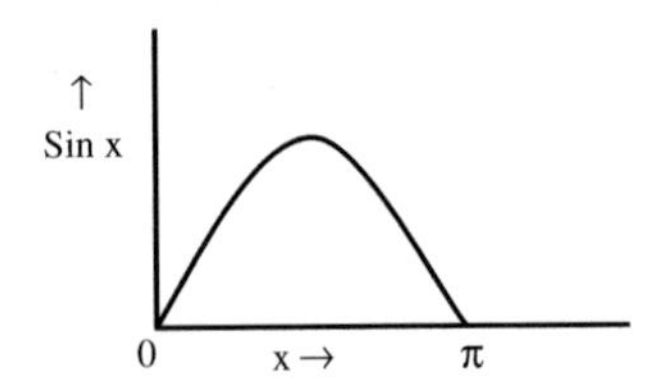


**Fig. 2.14** Figure for Problem 4

[**Hints:** The inference generation is examined below. Add figure given in Fig. 2.16]

6. In stock index prediction, we use the secondary factor, here DOW JONES stock index data for the main factor TAIEX time-series as indicated in Fig. 2.17. On the day (t − 1), it is observed that the secondary index lies in partition $B_5$, while the main factor time-series falls in $A_3$. Given the rules under group $B_5$ (Fig. 2.18):

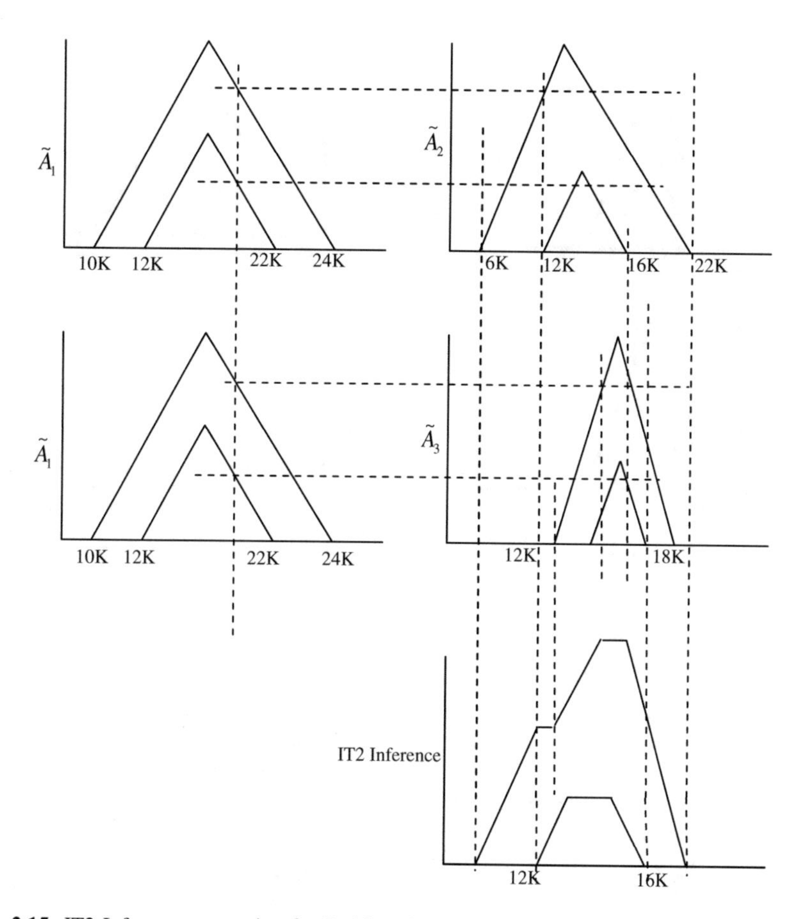


**Fig. 2.15** IT2 Inference generation for Problem 4

$$
\begin{aligned}
&A_1 \rightarrow \\
&A_2 \rightarrow \\
&A_3 \rightarrow A_4, A_6 \\
&A_4 \rightarrow \\
&\ldots \\
&A_{5\rightarrow}
\end{aligned}
$$

Given the MFs of $A_3, A_4, A_6$ as follows and today's price as 1100. Determine the fuzzy inference (Fig. 2.19).

7. A close price time-series is partitioned into 4 partitions: $P_1$, $P_2$, $P_3$ and $P_4$. The close price falling in $P_i$ would have a membership 1 in $A_i$ fuzzy sets and membership 0.5 in $A_{i-1}$ and $A_{i+1}$ and zero elsewhere. If the range of $P_1$, $P_2$, $P_3$ and $P_4$ are [0, 1 K), [1 K, 2 K), [2 K, 3 K) and [3 K, 4 K] respectively, construct $A_1$, $A_2$, $A_3$ and $A_4$ (Fig. 2.20).

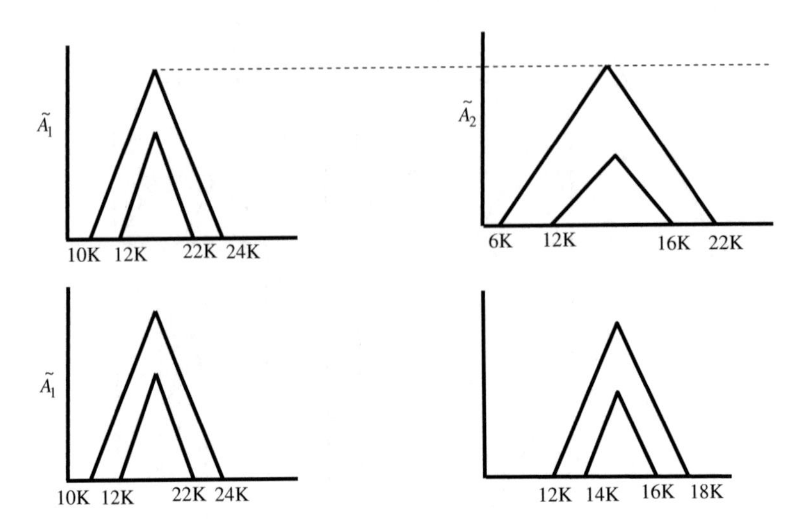


**Fig. 2.16** Figure for Problem 5

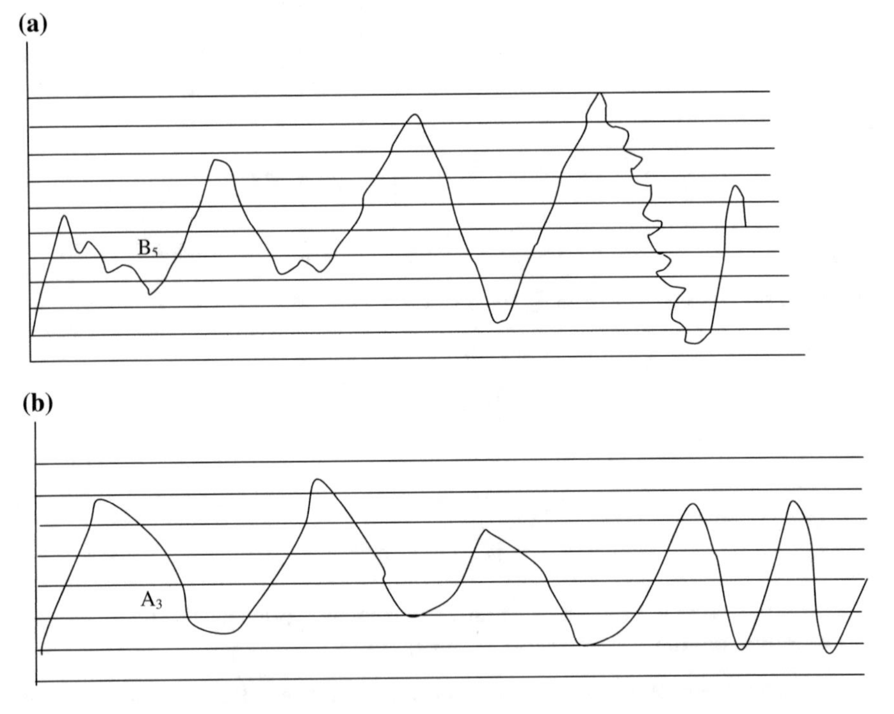


**Fig. 2.17** Figure for Problem 6

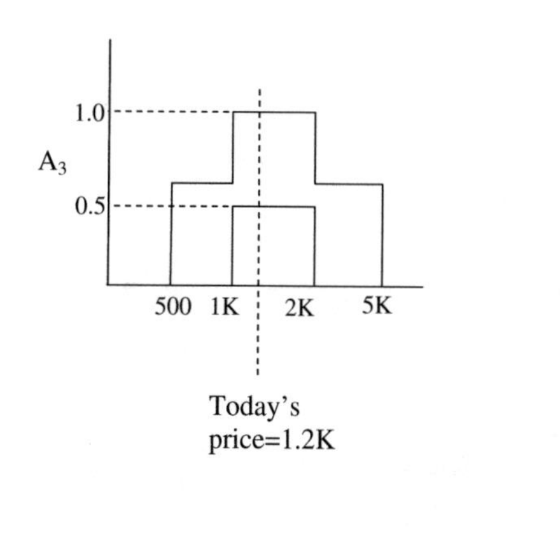


**Fig. 2.18** Figure for Problem 6

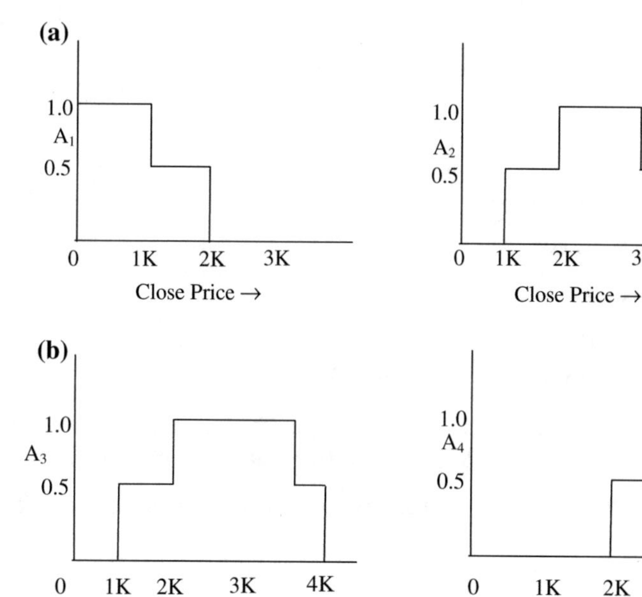


**Fig. 2.19** Figure for Problem 7

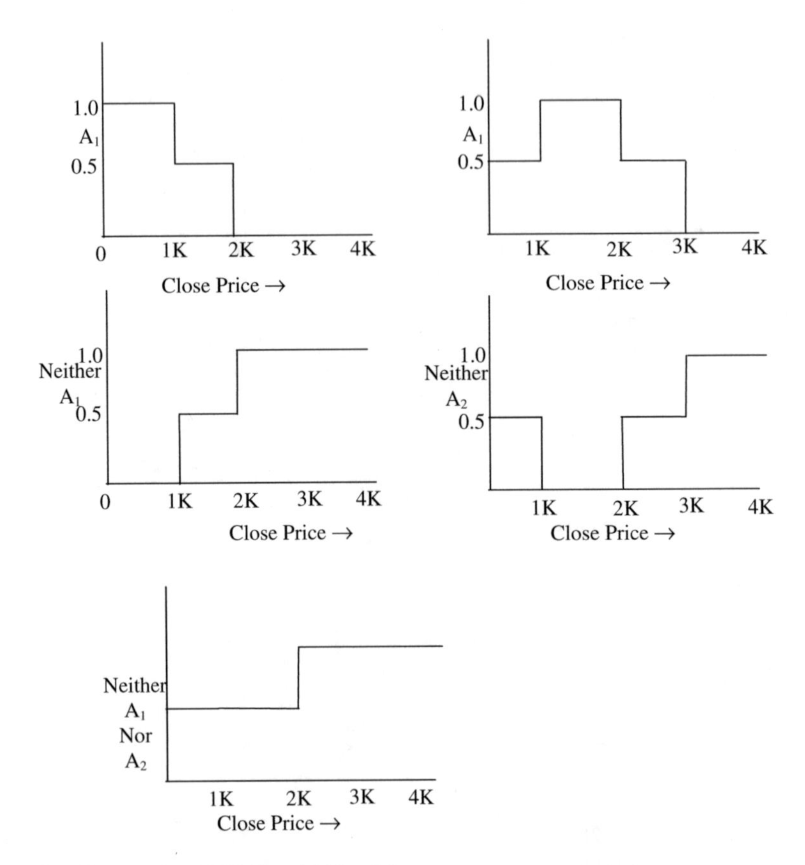


**Fig. 2.20** Computation of Neither A1 Nor A2

[**Hints**: membership functions are given below.]

8. In question 7 suppose we need to construct the membership function of (i) neither $A_1$ nor $A_2$, (ii) either $A_1$ or $A_3$, (iii) neither $A_1$ nor $A_2$ and $A_3$.

   [**Hints**: We show the solution for part (i). The rest can be obtained similarly.]

## Appendix 2.1

See Tables 2.6, 2.7, 2.8, 2.9, 2.10, 2.11, 2.12 and 2.13.

**Table 2.7** Fuzzy logical implication rules for year 2004

| Group | Fuzzy logical implication |
|---|---|
| $B_3$ | $A_3 \rightarrow A_2$ |
| $B_4$ | $A_2 \rightarrow A_1, A_2$; $A_3 \rightarrow A_2$; $A_5 \rightarrow A_4$ |
| $B_5$ | $A_1 \rightarrow A_1$; $A_4 \rightarrow A_4, A_5$; $A_2 \rightarrow A_2$;<br>$A_6 \rightarrow A_6, A_7$; $A_3 \rightarrow A_3$;$A_8 \rightarrow A_8$ |
| $B_6$ | $A_1 \rightarrow A_1, A_1$; $A_7 \rightarrow A_6$;<br>$A_2 \rightarrow A_1, A_1, A_2, A_2$;$A_8 \rightarrow A_8$;<br>$A_3 \rightarrow A_2, A_2, A_3, A_3$;<br>$A_9 \rightarrow A_8$; $A_4 \rightarrow A_3$; $A_{10} \rightarrow A_{10}$;<br>$A_5 \rightarrow A_5, A_5, A_5$;$A_{11} \rightarrow A_{10}$;$A_6 \rightarrow A_6$ |
| $B_7$ | $A_1 \rightarrow A_1, A_1, A_1$; $A_7 \rightarrow A_7, A_7, A_7, A_7, A_7, A_7$;<br>$A_2 \rightarrow A_2, A_2, A_2, A_2, A_2, A_2, A_2$;<br>$A_8 \rightarrow A_8, A_8, A_8, A_8, A_8, A_8, A_8, A_8, A_8$;<br>$A_3 \rightarrow A_2, A_2, A_2, A_3, A_3, A_3, A_3, A_3, A_3, A_3, A_3, A_3, A_3, A_3, A_3$;<br>$A_4 \rightarrow A_4, A_4, A_4, A_4$;$A_9 \rightarrow A_8, A_8, A_8, A_9$;<br>$A_5 \rightarrow A_4, A_5, A_5, A_5, A_5, A_5$; $A_{10} \rightarrow A_{10}, A_{10}, A_{10}, A_{10}, A_{10}$; |
| $B_8$ | $A_1 \rightarrow A_1, A_1, A_1, A_1, A_1$; $A_6 \rightarrow A_6, A_6, A_6, A_6$;<br>$A_2 \rightarrow A_2, A_2, A_2, A_2, A_2, A_2, A_3$; $A_7 \rightarrow A_7, A_7, A_7, A_8$;<br>$A_3 \rightarrow A_3, A_3, A_3, A_3, A_3, A_4$;<br>$A_8 \rightarrow A_8, A_8, A_8, A_8, A_8, A_8, A_8, A_9, A_9, A_9, A_9$;<br>$A_4 \rightarrow A_4, A_4, A_4, A_4, A_4$; $A_9 \rightarrow A_9, A_9, A_9, A_9, A_9$;<br>$A_5 \rightarrow A_5, A_5, A_5, A_5, A_5, A_5, A_5, A_5, A_5$; |
| $B_9$ | $A_1 \rightarrow A_1, A_1, A_1, A_1, A_2, A_2, A_2$;<br>$A_7 \rightarrow A_7, A_7, A_7$;$A_2 \rightarrow A_2, A_2, A_3, A_3$;<br>$A_8 \rightarrow A_8, A_8, A_9$;$A_3 \rightarrow A_3, A_3, A_3, A_4$;$A_9 \rightarrow A_9, A_{10}$;<br>$A_4 \rightarrow A_4, A_4, A_5$;$A_{10} \rightarrow A_{10}, A_{10}$;$A_5 \rightarrow A_5, A_5, A_6$ |
| $B_{10}$ | $A_2 \rightarrow A_2, A_2, A_3, A_3$;$A_7 \rightarrow A_7, A_7$;$A_3 \rightarrow A_3, A_3$;<br>$A_8 \rightarrow A_9, A_9$;$A_4 \rightarrow A_4, A_5, A_5$ |
| $B_{11}$ | $A_2 \rightarrow A_2$;$A_6 \rightarrow A_7, A_7$;$A_3 \rightarrow A_3$ |
| $B_{12}$ | $A_2 \rightarrow A_3, A_4$ |

**Table 2.8** Fuzzy logical implication rules considering d = 2

| Group | Fuzzy logical relationship |
|---|---|
| $B_1$ | $A_3 \rightarrow A_2$; $A_6 \rightarrow A_4$, $A_5$; $A_8 \rightarrow A_5$; |
| $B_2$ | $A_3 \rightarrow A_2, A_3$; $A_5 \rightarrow A_5$; |
| $B_3$ | $A_3 \rightarrow A_4$; |
| $B_4$ | $A_1 \rightarrow A_1, A_3$;$A_3 \rightarrow A_2, A_3$; $A_4 \rightarrow A_1$; $A_7 \rightarrow A_5, A_7$; $A_6 \rightarrow A_7$; |
| $B_5$ | $A_2 \rightarrow A_1, A_2, A_2, A_2, A_4$;$A_3 \rightarrow A_3, A_4A_4 \rightarrow A_3, A_3, A_3, A_4$;<br>$A_5 \rightarrow A_4, A_5$; $A_7 \rightarrow A_6$; |

(continued)

**Table 2.8** (continued)

| Group | Fuzzy logical relationship |
|---|---|
| $B_6$ | $A_1 \to A_1, A_1; A_2 \to A_2, A_2; A_3 \to A_2, A_2, A_2, A_3, A_3, A_3, A_3, A_3, A_3;$ $A_4 \to A_2, A_3, A_3, A_3; A_5 \to A_5, A_6; A_6 \to A_5, A_6; A_7 \to A_7;$ |
| $B_7$ | $A_1 \to A_1, A_1, A_1, A_1, A_1, A_1, A_1, A_1, A_1, A_1;$ $A_2 \to A_1, A_2, A_3, A_3;$ $A_3 \to A_2, A_3, A_3, A_3, A_3, A_3, A_3, A_3, A_3, A_3, A_3, A_3;$ $A_4 \to A_3, A_3, A_4, A_5; A_5 \to A_4, A_4, A_5, A_5, A_5;$ $A_6 \to A_6, A_6, A_7; A_7 \to A_7, A_7, A_7, A_8, A_8; A_6 \to A_6, A_6, A_7;$ $A_7 \to A_7, A_7, A_7, A_8, A_8; A_8 \to A_7, A_8, A_8, A_8, A_8, A_8;$ $A_9 \to A_8$ |
| $B_8$ | $A_1 \to A_1, A_1, A_1, A_1, A_1; A_2 \to A_2, A_2, A_2, A_3, A_3; A_8 \to A_8, A_8$ $A_3 \to A_2, A_3, A_3, A_4; A_4 \to A_3, A_3, A_5, A_5; A_9 \to A_8, A_9, A_9, A_9$ $A_5 \to A_4, A_5, A_5, A_5, A_5, A_5, A_5, A_5, A_5, A_5, A_6, A_6$ $A_6 \to A_5, A_7; A_7 \to A_7, A_7, A_7, A_7, A_7, A_7;$ |
| $B_9$ | $A_1 \to A_1, A_1, A_1, A_2; A_2 \to A_2; A_3 \to A_3, A_3, A_3, A_3, A_3, A_3, A_3, A_3, A_4, A_4$ $A_4 \to A_3, A_3, A_4, A_4, A_4, A_5; A_5 \to A_5, A_5, A_6, A_6;$ $A_6 \to A_6, A_7; A_7 \to A_7, A_7, A_7, A_8, A_8, A_8; A_8 \to A_8$ |
| $B_{10}$ | $A_2 \to A_3; A_3 \to A_3, A_3, A_4, A_4, A_4; A_4 \to A_4, A_4; A_6 \to A_6; A_7 \to A_7, A_8; A_8 \to A_8, A_9;$ |
| $B_{11}$ | $A_2 \to A_2, A_3; A_4 \to A_3, A_4; A_6 \to A_6; A_8 \to A_6$ |
| $B_{12}$ | $A_1 \to A_2; A_3 \to A_3, A_4, A_4; A_4 \to A_4$ |
| $B_{13}$ | $A_5 \to A_6; A_6 \to A_7;$ |
| $B_{14}$ | $A_2 \to A_3;$ |

**Table 2.9** Fuzzy logical implication rules considering d = 3

| Group | Fuzzy logical relationship |
|---|---|
| $B_1$ | $A_1 \to A_3; A_3 \to A_2, A_3;$ $A_5 \to A_3, A_5; A_6 \to A_5, A_5;$ |
| $B_2$ | $A_3 \to A_4;$ |
| $B_3$ | $A_2 \to A_4; A_3 \to A_2; A_4 \to A_3, A_3;$ $A_5 \to A_4; A_7 \to A_4, A_8$ |
| $B_4$ | $A_1 \to A_1; A_7 \to A_8; A_2 \to A_1, A_1, A_2, A_2; A_8 \to A_5, A_7;$ $A_3 \to A_3, A_3; A_4 \to A_4; A_5 \to A_6;$ |
| $B_5$ | $A_2 \to A_1, A_2, A_3; A_3 \to A_2, A_3, A_3;$ $A_4 \to A_3, A_3, A_3; A_5 \to A_6; A_7 \to A_5; A_8 \to A_7;$ |
| $B_6$ | $A_1 \to A_1, A_1, A_1, A_1; A_2 \to A_2, A_3, A_3;$ $A_3 \to A_3, A_3, A_3, A_3, A_4, A_4;$ $A_4 \to A_2, A_2; A_6 \to A_5; A_7 \to A_7; A_8 \to A_6, A_7, A_8;$ |
| $B_7$ | $A_1 \to A_1, A_1, A_1, A_1, A_1, A_1; A_2 \to A_3;$ $A_3 \to A_2, A_2, A_2, A_3, A_3, A_3, A_3, A_3, A_3, A_3, A_3, A_3, A_4, A_4, A_4, A_4;$ $A_4 \to A_2, A_3, A_4; A_5 \to A_3, A_5, A_6, A_6;$ $A_6 \to A_5, A_5, A_7;$ $A_7 \to A_8, A_8, A_8; A_8 \to A_7, A_7, A_8, A_8, A_8; A_9 \to A_8, A_8, A_8, A_9;$ |
| $B_8$ | $A_1 \to A_1, A_1, A_1, A_1, A_1, A_1, A_1, A_2;$ $A_2 \to A_3;$ $A_3 \to A_2, A_2, A_3, A_3, A_3, A_3, A_4;$ |

(continued)

**Table 2.9** (continued)

| Group | Fuzzy logical relationship |
|---|---|
| | $A_4 \to A_3, A_4$;<br>$A_5 \to A_4, A_4, A_5, A_5, A_5, A_5, A_5, A_5, A_5, A_6, A_6$;<br>$A_6 \to A_6, A_7; A_7 \to A_7, A_7, A_7, A_7, A_7, A_8$;<br>$A_8 \to A_8, A_8$; |
| $B_9$ | $A_1 \to A_1, A_1; A_2 \to A_2, A_3, A_4$;<br>$A_3 \to A_3, A_3, A_3, A_3, A_4, A_4, A_4$;<br>$A_4 \to A_1, A_3, A_3, A_4, A_4, A_5, A_5$;<br>$A_5 \to A_5, A_5, A_5, A_6, A_6$;<br>$A_6 \to A_6, A_7; A_7 \to A_7, A_7, A_7, A_7, A_8; A_8 \to A_8$; |
| $B_{10}$ | $A_2 \to A_3, A_3; A_3 \to A_2, A_3, A_3; A_4 \to A_3, A_4$;<br>$A_7 \to A_6, A_7, A_7, A_9; A_9 \to A_9, A_9$; |
| $B_{11}$ | $A_4 \to A_4; A_6 \to A_6, A_6$; |
| $B_{12}$ | $A_1 \to A_2; A_2 \to A_2; A_3 \to A_3, A_3$;<br>$A_4 \to A_3, A_4, A_5; A_5 \to A_7; A_8 \to A_9$; |
| $B_{13}$ | $A_2 \to A_2; A_4 \to A_3; A_6 \to A_7$; |
| $B_{14}$ | $A_3 \to A_4, A_4; A_6 \to A_7$; |

**Table 2.10** Fuzzy logical implication rules considering d = 4

| Group | Fuzzy logical relationship |
|---|---|
| $B_1$ | $A_1 \to A_4$; $A_3 \to A_2, A_3, A_4$; $A_5 \to A_3, A_4, A_6, A_6$; $A_6 \to A_3, A_5$ |
| $B_2$ | $A_1 \to A_1$; $A_2 \to A_2, A_2$; $A_4 \to A_4, A_4$; $A_7 \to A_5$ |
| $B_3$ | $A_2 \to A_1, A_1, A_4$; $A_3 \to A_3$; $A_4 \to A_3$; $A_7 \to A_6, A_8$; $A_8 \to A_8$ |
| $B_4$ | $A_2 \to, A_1$; $A_8 \to A_5$; $A_3 \to A_2, A_3, A_3$; $A_5 \to A_7$; |
| $B_5$ | $A_1 \to A_1, A_1, A_1$; $A_3 \to A_3, A_3, A_4, A_4$; $A_4 \to A_2, A_3, A_4$;<br>$A_6 \to A_6$; $A_7 \to A_4$; $A_8 \to A_5$; $A_9 \to A_7$; |
| $B_6$ | $A_1 \to A_1$; $A_2 \to A_1, A_2, A_3, A_3$; $A_3 \to A_3, A_3, A_3, A_4$;<br>$A_4 \to A_3$; $A_5 \to A_6, A_6$; $A_7 \to A_5$; $A_8 \to A_6, A_7, A_7, A_8$; $A_9 \to A_8$; |
| $B_7$ | $A_1 \to A_1, A_1, A_1, A_1, A_1, A_1, A_1, A_1, A_2, A_2$; $A_2 \to A_3, A_3, A_4$;<br>$A_3 \to A_2, A_2, A_2, A_2, A_3, A_3, A_3, A_3, A_4, A_4, A_4$;<br>$A_4 \to A_3, A_3$; $A_5 \to A_5$; $A_7 \to A_5, A_8, A_8, A_8$; $A_8 \to A_7, A_7, A_8$; |
| $B_8$ | $A_1 \to A_1, A_1, A_2$; $A_2 \to A_3, A_3$; $A_3 \to A_2, A_3, A_3, A_3, A_3, A_4, A_4$;<br>$A_4 \to A_3, A_3, A_3, A_4, A_5$; $A_5 \to A_5, A_5, A_5, A_5, A_6$;<br>$A_6 \to A_5, A_6, A_7, A_7$; $A_7 \to A_7, A_7, A_7, A_7, A_8, A_8, A_9$;<br>$A_8 \to A_8, A_8$; |
| $B_9$ | $A_1 \to A_1, A_1$; $A_2 \to A_2, A_3, A_4$; $A_3 \to A_3, A_3, A_3, A_3, A_4, A_4, A_4$;<br>$A_4 \to A_1, A_3, A_3, A_4, A_4, A_5, A_5$; $A_5 \to A_5, A_5, A_5, A_6, A_6$;<br>$A_6 \to A_6, A_7$; $A_7 \to A_7, A_7, A_7, A_7, A_8$; $A_8 \to A_8$; |
| $B_{10}$ | $A_3 \to A_2, A_3, A_3, A_3$; $A_4 \to A_2, A_3, A_4$; $A_5 \to A_5$; $A_6 \to A_7$;<br>$A_7 \to A_7, A_7, A_7, A_7, A_8$; $A_8 \to A_5, A_8$; |
| $B_{11}$ | $A_1 \to A_2$; $A_2 \to A_3$; $A_3 \to A_3$; $A_4 \to A_3, A_4, A_4, A_5$;<br>$A_6 \to A_6, A_7$; $A_8 \to A_8$; $A_9 \to A_9$; |
| $B_{12}$ | $A_2 \to A_3$; $A_3 \to A_3$; $A_4 \to A_5$; $A_6 \to A_7$; $A_8 \to A_9, A_9$; $A_9 \to A_9$; |
| $B_{13}$ | $A_2 \to A_2, A_3$; $A_4 \to A_2$; $A_6 \to A_7$; |
| $B_{14}$ | $A_3 \to A_4$; $A_4 \to A_4$; $A_7 \to A_7$ |

**Table 2.11** Frequency of occurrence of main factor variation in each group of secondary factor variation considering d = 1

| | | Main factor variation | | | | | | | | | | | | | |
|---|---|---|---|---|---|---|---|---|---|---|---|---|---|---|---|
| | | $B_1$ | $B_2$ | $B_3$ | $B_4$ | $B_5$ | $B_6$ | $B_7$ | $B_8$ | $B_9$ | $B_{10}$ | $B_{11}$ | $B_{12}$ | $B_{13}$ | $B_{14}$ |
| Secondary factor variation | $B'_1$ | 0 | 0 | 0 | 0 | 0 | 0 | 0 | 0 | 0 | 0 | 0 | 0 | 0 | 0 |
| | $B'_2$ | 0 | 0 | 0 | 0 | 0 | 0 | 0 | 0 | 0 | 0 | 0 | 0 | 0 | 0 |
| | $B'_3$ | 0 | 0 | 0 | 0 | 0 | 0 | 0 | 0 | 0 | 0 | 0 | 0 | 0 | 0 |
| | $B'_4$ | 0 | 0 | 0 | 0 | 0 | 2 | 0 | 0 | 0 | 0 | 0 | 0 | 0 | 0 |
| | $B'_5$ | 0 | 0 | 0 | 0 | 1 | 1 | 2 | 1 | 0 | 0 | 0 | 0 | 0 | 0 |
| | $B'_6$ | 0 | 0 | 0 | 1 | 2 | 5 | 9 | 10 | 0 | 0 | 0 | 0 | 0 | 0 |
| | $B'_7$ | 0 | 0 | 0 | 2 | 2 | 7 | 19 | 13 | 11 | 1 | 1 | 0 | 0 | 0 |
| | $B'_8$ | 0 | 0 | 1 | 0 | 4 | 3 | 24 | 23 | 10 | 8 | 1 | 0 | 0 | 0 |
| | $B'_9$ | 0 | 0 | 0 | 1 | 0 | 1 | 9 | 10 | 8 | 1 | 2 | 0 | 0 | 0 |
| | $B'_{10}$ | 0 | 0 | 0 | 0 | 0 | 1 | 1 | 4 | 1 | 1 | 0 | 1 | 0 | 0 |
| | $B'_{11}$ | 0 | 0 | 0 | 0 | 0 | 0 | 0 | 0 | 1 | 0 | 0 | 1 | 0 | 0 |
| | $B'_{12}$ | 0 | 0 | 0 | 0 | 0 | 0 | 0 | 0 | 0 | 1 | 0 | 0 | 0 | 0 |
| | $B'_{13}$ | 0 | 0 | 0 | 0 | 0 | 0 | 0 | 0 | 0 | 0 | 0 | 0 | 0 | 0 |
| | $B'_{14}$ | 0 | 0 | 0 | 0 | 0 | 0 | 0 | 0 | 0 | 0 | 0 | 0 | 0 | 0 |

**Table 2.12** Frequency of occurrence of main factor variation in each group of secondary factor variation considering d = 2

| | | Main factor variation | | | | | | | | | | | | | |
|---|---|---|---|---|---|---|---|---|---|---|---|---|---|---|---|
| | | $B_1$ | $B_2$ | $B_3$ | $B_4$ | $B_5$ | $B_6$ | $B_7$ | $B_8$ | $B_9$ | $B_{10}$ | $B_{11}$ | $B_{12}$ | $B_{13}$ | $B_{14}$ |
| Secondary factor variation | $B'_1$ | 0 | 0 | 0 | 0 | 0 | 0 | 0 | 0 | 0 | 0 | 0 | 0 | 0 | 0 |
| | $B'_2$ | 0 | 0 | 0 | 0 | 0 | 0 | 0 | 0 | 0 | 0 | 0 | 0 | 0 | 0 |
| | $B'_3$ | 0 | 0 | 0 | 0 | 0 | 0 | 0 | 0 | 0 | 0 | 0 | 0 | 0 | 0 |
| | $B'_4$ | 0 | 0 | 0 | 0 | 0 | 0 | 0 | 0 | 0 | 0 | 0 | 0 | 0 | 0 |
| | $B'_5$ | 0 | 0 | 0 | 0 | 0 | 0 | 0 | 0 | 0 | 0 | 0 | 0 | 0 | 0 |
| | $B'_6$ | 2 | 1 | 0 | 3 | 3 | 4 | 12 | 2 | 4 | 1 | 0 | 1 | 0 | 1 |
| | $B'_7$ | 2 | 0 | 0 | 3 | 8 | 11 | 15 | 18 | 9 | 2 | 3 | 0 | 1 | 0 |
| | $B'_8$ | 0 | 2 | 1 | 1 | 3 | 10 | 22 | 17 | 12 | 7 | 2 | 4 | 0 | 0 |
| | $B'_9$ | 0 | 0 | 0 | 1 | 0 | 1 | 3 | 7 | 7 | 3 | 2 | 0 | 0 | 0 |
| | $B'_{10}$ | 0 | 0 | 0 | 0 | 0 | 0 | 0 | 0 | 0 | 0 | 0 | 0 | 1 | 0 |
| | $B'_{11}$ | 0 | 0 | 0 | 0 | 0 | 0 | 0 | 0 | 0 | 0 | 0 | 0 | 0 | 0 |
| | $B'_{12}$ | 0 | 0 | 0 | 0 | 0 | 0 | 0 | 0 | 0 | 0 | 0 | 0 | 0 | 0 |
| | $B'_{13}$ | 0 | 0 | 0 | 0 | 0 | 0 | 0 | 0 | 0 | 0 | 0 | 0 | 0 | 0 |
| | $B'_{14}$ | 0 | 0 | 0 | 0 | 0 | 0 | 0 | 0 | 0 | 0 | 0 | 0 | 0 | 0 |

**Table 2.13** Frequency of occurrence of main factor variation in each group of secondary factor variation considering d = 3

| | | Main factor variation | | | | | | | | | | | | | |
|---|---|---|---|---|---|---|---|---|---|---|---|---|---|---|---|
| | | $B_1$ | $B_2$ | $B_3$ | $B_4$ | $B_5$ | $B_6$ | $B_7$ | $B_8$ | $B_9$ | $B_{10}$ | $B_{11}$ | $B_{12}$ | $B_{13}$ | $B_{14}$ |
| Secondary factor variation | $B'_1$ | 0 | 0 | 0 | 0 | 0 | 0 | 0 | 0 | 0 | 0 | 0 | 0 | 0 | 0 |
| | $B'_2$ | 0 | 0 | 0 | 0 | 0 | 0 | 0 | 0 | 0 | 0 | 0 | 0 | 0 | 0 |
| | $B'_3$ | 0 | 0 | 0 | 0 | 0 | 0 | 0 | 0 | 0 | 0 | 0 | 0 | 0 | 0 |
| | $B'_4$ | 0 | 0 | 0 | 0 | 0 | 0 | 0 | 0 | 0 | 0 | 0 | 0 | 0 | 0 |
| | $B'_5$ | 1 | 0 | 0 | 0 | 0 | 1 | 2 | 2 | 0 | 1 | 1 | 0 | 0 | 0 |
| | $B'_6$ | 2 | 0 | 2 | 3 | 4 | 5 | 7 | 4 | 6 | 0 | 0 | 0 | 1 | 1 |
| | $B'_7$ | 3 | 0 | 1 | 4 | 4 | 4 | 19 | 7 | 6 | 7 | 0 | 1 | 0 | 1 |
| | $B'_8$ | 1 | 1 | 4 | 3 | 3 | 7 | 12 | 15 | 14 | 3 | 4 | 4 | 1 | 0 |
| | $B'_9$ | 0 | 0 | 0 | 1 | 1 | 3 | 6 | 11 | 2 | 2 | 0 | 3 | 0 | 1 |
| | $B'_{10}$ | 0 | 0 | 0 | 1 | 0 | 1 | 0 | 1 | 3 | 0 | 1 | 1 | 0 | 0 |
| | $B'_{11}$ | 0 | 0 | 0 | 0 | 0 | 0 | 0 | 0 | 0 | 0 | 0 | 0 | 1 | 0 |
| | $B'_{12}$ | 0 | 0 | 0 | 0 | 0 | 0 | 0 | 0 | 0 | 0 | 0 | 0 | 0 | 0 |
| | $B'_{13}$ | 0 | 0 | 0 | 0 | 0 | 0 | 0 | 0 | 0 | 0 | 0 | 0 | 0 | 0 |
| | $B'_{14}$ | 0 | 0 | 0 | 0 | 0 | 0 | 0 | 0 | 0 | 0 | 0 | 0 | 0 | 0 |

**Table 2.14** Frequency of occurrence of main factor variation in each group of secondary factor variation considering d = 4

| | | Main factor variation | | | | | | | | | | | | | |
|---|---|---|---|---|---|---|---|---|---|---|---|---|---|---|---|
| | | $B_1$ | $B_2$ | $B_3$ | $B_4$ | $B_5$ | $B_6$ | $B_7$ | $B_8$ | $B_9$ | $B_{10}$ | $B_{11}$ | $B_{12}$ | $B_{13}$ | $B_{14}$ |
| Secondary factor variation | $B'_1$ | 0 | 0 | 0 | 0 | 0 | 0 | 0 | 0 | 0 | 0 | 0 | 0 | 0 | 0 |
| | $B'_2$ | 0 | 0 | 0 | 0 | 0 | 0 | 0 | 0 | 0 | 0 | 0 | 0 | 0 | 0 |
| | $B'_3$ | 0 | 0 | 0 | 0 | 0 | 0 | 0 | 0 | 0 | 0 | 0 | 0 | 0 | 0 |
| | $B'_4$ | 0 | 0 | 0 | 0 | 0 | 0 | 0 | 0 | 0 | 0 | 0 | 0 | 0 | 0 |
| | $B'_5$ | 0 | 0 | 0 | 0 | 0 | 0 | 0 | 0 | 0 | 0 | 0 | 0 | 0 | 0 |
| | $B'_6$ | 2 | 2 | 3 | 2 | 4 | 7 | 7 | 2 | 1 | 1 | 1 | 0 | 1 | 1 |
| | $B'_7$ | 4 | 1 | 2 | 2 | 5 | 5 | 11 | 14 | 15 | 5 | 3 | 4 | 0 | 1 |
| | $B'_8$ | 2 | 3 | 3 | 1 | 5 | 8 | 12 | 14 | 13 | 8 | 7 | 3 | 1 | 1 |
| | $B'_9$ | 2 | 0 | 0 | 1 | 1 | 0 | 4 | 5 | 6 | 1 | 0 | 1 | 1 | 0 |
| | $B'_{10}$ | 0 | 0 | 0 | 0 | 0 | 0 | 0 | 0 | 0 | 0 | 0 | 0 | 1 | 0 |
| | $B'_{11}$ | 0 | 0 | 0 | 0 | 0 | 0 | 0 | 0 | 0 | 0 | 0 | 0 | 0 | 0 |
| | $B'_{12}$ | 0 | 0 | 0 | 0 | 0 | 0 | 0 | 0 | 0 | 0 | 0 | 0 | 0 | 0 |
| | $B'_{13}$ | 0 | 0 | 0 | 0 | 0 | 0 | 0 | 0 | 0 | 0 | 0 | 0 | 0 | 0 |
| | $B'_{14}$ | 0 | 0 | 0 | 0 | 0 | 0 | 0 | 0 | 0 | 0 | 0 | 0 | 0 | 0 |

## Appendix 2.2: Source Codes of the Programs

% MATLAB Source Code of the Main Program and Other Functions for Time
% Series Prediction by IT2FS REasoning

% Developed by Monalisa Pal

% Under the guidance of Amit Konar and Diptendu Bhattacharya

```
% Main Program
%______________________________________________________________
clear all;
clc;
str=input('Data file having Main Factor:');
load(str);
closeMF=close;
n=input('Number of secondary factors:');
n=str2double(n);
SF=zeros(length(closeMF),n);
for i=1:n
   strSF=input(['Data file having ',num2str(i),'-th    Secondary Factor:']);
   load(strSF);
   SF(:,i)=close;
end
clear close;
tic;
%% Training
[A,B,VarMF,Au,Al]=partitioning(closeMF);
plotpartitions(closeMF(s:e1),A(s:e1),Al,Au);
FLRG=tableVI(B(2:e1),A(1:e1-1),A(2:e1));
VarSF=overallvar(VarMF,SF,e1,n);
BSF=fuzzyvarSF(VarSF);
FVG=tableIX(B(2:e1),BSF(2:e1));
WBS=tableX(FVG);
%% Validation (Differential Evolution)
sd=extractSD(A(s:e1),closeMF(s:e1));
[UMF,LMF,typeMF,rmse]=MFusingDE(A(s:e1),closeMF(s:e1),
Al:Au-1,sd,closeMF(e1+1:e2),A(e1:e2),FLRG,WBS,BSF(e1:e2));
plotDE(rmse);
plotFOU(UMF,LMF,Al:Au-1);
%% Inference
forecasted=predict(closeMF(e2+1:f),A(e2:f),Al,Au,UMF,LMF,typeMF,
FLRG,WBS,BSF(e2:f));
[CFE,ME,MSE,RMSE,SD,MAD,MAPE]=
errormetrics(forecasted,closeMF(e2+1:f));
```

```
disp('CFE');disp(CFE);
disp('ME');disp(ME);
disp('MSE');disp(MSE);
disp('RMSE');disp(RMSE);
disp('SD');disp(SD);
disp('MAD');disp(MAD);
disp('MAPE');disp(MAPE);
plotforecasted(forecasted,closeMF(e2+1:f),RMSE);
comp_time=toc;
disp('Execution Time');disp(comp_time);
%End of Main Program

% Function KMmethod to compute IT2FS Centroid
%_______________________________________________

function centroid=KMmethod(LMF,UMF,x)

diff=ones(1,length(x))*5000;
for i=1:length(x)
   if UMF(i)>0
      theta(i,:)=[UMF(1:i) LMF(i+1:length(LMF))];
      centroid(i)=defuzz(x,theta(i,:),'centroid');
      diff(i)=abs(x(i)-centroid(i));
   end
end
[mindiff sw_index]=min(diff);
cl=x(sw_index);
theta_l=[UMF(1:sw_index) LMF(sw_index+1:length(LMF))];

diff=ones(1,length(x))*5000;
for i=1:length(x);
   if  i<length(x)
      if UMF(i+1)>0
         theta(i,:)=[LMF(1:i) UMF(i+1:length(LMF))];
         centroid(i)=defuzz(x,theta(i,:),'centroid');
         diff(i)=abs(x(i)-centroid(i));
      end
   end
end;
[mindiff sw_index]=min(diff);
cr=x(sw_index);
theta_r=[LMF(1:sw_index) UMF(sw_index+1:length(LMF))];
```

```
centroid=(cl+cr)/2;
end
% End of functiom KMmethod

%%%%%%%%%%%%%%%%%%%%%%%%%%
% Error Metric Calculation

function [CFE,ME,MSE,RMSE,SD,MAD,MAPE]=errormetrics(predicted,TestCP)
% a=isnan(predicted1);
% p=size(TestCP1);
% z=1;
% for i=1:p
%   if a(i)~=1
%     predicted(z)=predicted1(i);
%     TestCP(z)=TestCP1(i);z=z+1;
%   end
% end
  CFE=sum(TestCP-predicted);
%   disp('CFE=');
%   disp(CFE);
  ME=mean(TestCP-predicted);
%   disp('ME=');
%   disp(ME);
  MSE=mean((TestCP-predicted).^2);
%   disp('MSE=');
%   disp(MSE);
  RMSE=sqrt(MSE);
%   disp('RMSE=');
%   disp(RMSE);
  SD=std(TestCP-predicted);
%   disp('SD=');
%   disp(SD);
  MAD=mean(abs(TestCP-predicted));
%   disp('MAD=');
%   disp(MAD);
  MAPE=mean(abs(TestCP-predicted)./TestCP)*100;
%   disp('MAPE=');
%   disp(MAPE);
End

%%%%%%%%%%%%%%%%
% RMSE calculation

function rmse=evalfit(x,TestCP,TestA,UMF,LMF,typeMF,FLRG,WBS,TestB)
  Al=x(1);
```

```
   Au=x(end)+1;
          forecasted=predict(TestCP,TestA,Al,Au,UMF,LMF,typeMF,FLRG,WBS,
TestB);
   [~,~,~,rmse,~,~,~]=errormetrics(forecasted,TestCP);
End

%%%%%%%%%%%%%%%%
% Find Standard Deviations(SD)

function sd=extractSD(A,close)
   numFS=unique(A); % converting the close series into a matrix
   FS=zeros(length(numFS),length(close));
   for i=1:length(A)
      FS(A(i),i)=close(i);
   end
   b=1;
   for i=1:size(FS,1)
      temp=find(FS(i,:));
      %Case2: Partition has one point
      if length(temp)==1
         sd(b)=0.001;
%          flag(b)=2;
         b=b+1;
      %Case3: Partition has two points
      elseif length(temp)==2
         sd(b)=std(close(temp));
%          flag(b)=3;
         b=b+1;
      %Case 1 and 4: More than 2 contiguous or discrete points
      else
         indx=zeros(length(temp),1);
         l=1;
         for j=2:length(temp)  %contiguous points have been labelled sequen-
tially
            if (temp(j)-temp(j-1))==1
               indx(j-1)=1;
               indx(j)=1;
            elseif j>2 && (temp(j-1)-temp(j-2))==1 && (temp(j)-temp(j-1))~=1
               l=l+1;
            end
         end
         if max(indx)==0
```

```
        sd(b)=std(close(temp));
%         flag(b)=4;
        b=b+1;
      else
        for j=1:max(indx)
              temp1=temp(indx==j); % selecting days where the contigu-
ous points occur
          tempsd=std(close(temp1));
%           if tempsd>=1
            sd(b)=tempsd;
%             flag(b)=1;
            b=b+1;
%           end
        end
      end
    end
  end
end
%%%%%%%%%%%%%%%%
% Forming Membership Functions(MFs)

function [UMF,LMF,typeMF]=formingMF(A,close,x,sd)
  numFS=unique(A); % converting the close series into a matrix
  FS=zeros(length(numFS),length(close));
  for i=1:length(A)
    FS(A(i),i)=close(i);
  end
  b=1;
  UMF=zeros(length(numFS),length(x));
  LMF=ones(length(numFS),length(x));
  typeMF=zeros(length(numFS),1);
  for i=1:size(FS,1)
    temp=find(FS(i,:));
    %Case2: Partition has one point
    if length(temp)==1
      m=round(close(temp));
      typeMF(i)=1;
      UMF(i,:)=max(UMF(i,:),trimf(x,[m-3*sd(b) m m+3*sd(b)]));
      LMF(i,:)=min(LMF(i,:),trimf(x,[m-3*sd(b) m m+3*sd(b)]));
      b=b+1;

    %Case3: Partition has two points
    elseif length(temp)==2
      m=round(mean(close(temp)));
```

```
        typeMF(i)=1;
        UMF(i,:)=max(UMF(i,:),trimf(x,[m-3*sd(b) m m+3*sd(b)]));
        LMF(i,:)=min(LMF(i,:),trimf(x,[m-3*sd(b) m m+3*sd(b)]));
        b=b+1;
    %Case 1 and 4: More than 2 contiguous or discrete points
    else
        indx=zeros(length(temp),1);
        l=1;
        for j=2:length(temp) %contiguous points have been labelled sequen-
tially
          if (temp(j)-temp(j-1))==1
            indx(j-1)=l;
            indx(j)=l;
         elseif j>2 && (temp(j-1)-temp(j-2))==1 && (temp(j)-temp(j-1))~=1
            l=l+1;
          end
        end
        if max(indx)==0
          m=round(mean(close(temp)));
          typeMF(i)=1;
          UMF(i,:)=max(UMF(i,:),trimf(x,[m-3*sd(b) m m+3*sd(b)]));
          LMF(i,:)=min(LMF(i,:),trimf(x,[m-3*sd(b) m m+3*sd(b)]));
          b=b+1;
        else
%           c=0;
          for j=1:max(indx)
                temp1=temp(indx==j); % selecting days where the contigu-
ous points occur
            m=round(mean(close(temp1)));
%              if sd>=1
              UMF(i,:)=max(UMF(i,:),trimf(x,[m-3*sd(b) m m+3*sd(b)]));
              LMF(i,:)=min(LMF(i,:),trimf(x,[m-3*sd(b) m m+3*sd(b)]));
              b=b+1;
%                c=c+1;
%              end
          end
%           if c==1
%              typeMF(i)=1;
%           else
            typeMF(i)=2;
%           end
        end
    end
  end
```

```
  %Ensuring flat top
  loc1=0;loc2=0;
  for i=1:size(UMF,1)
    for j=1:1:size(UMF,2)
      if UMF(i,j)>0.999
        loc1=j;
        break;
      end
    end
    if j~=size(UMF,2)
      for j=size(UMF,2):-1:1
        if UMF(i,j)>0.999
          loc2=j;
          break;
        end
      end
    end
    if loc1~=0
      UMF(i,loc1:loc2)=1;
    end
  end
end
%%%%%%%%%%%%%%%%%%%%%%%%%%%%%%%%
% Gaussian MF Creation

function [UMF,LMF,typeMF]=formingMF(A,close,x,sd)
  numFS=unique(A); % converting the close series into a matrix
  FS=zeros(length(numFS),length(close));
  for i=1:length(A)
    FS(A(i),i)=close(i);
  end
  b=1;
  UMF=zeros(length(numFS),length(x));
  LMF=ones(length(numFS),length(x));
  typeMF=zeros(length(numFS),1);
  for i=1:size(FS,1)
    temp=find(FS(i,:));
    %Case2: Partition has one point
    if length(temp)==1
      m=round(close(temp));
      typeMF(i)=1;
      UMF(i,:)=max(UMF(i,:),gaussmf(x,[sd(b) m]));
      LMF(i,:)=min(LMF(i,:),gaussmf(x,[sd(b) m]));
      b=b+1;
```

```
      %Case3: Partition has two points
      elseif length(temp)==2
        m=round(mean(close(temp)));
        typeMF(i)=1;
        UMF(i,:)=max(UMF(i,:),gaussmf(x,[sd(b) m]));
        LMF(i,:)=min(LMF(i,:),gaussmf(x,[sd(b) m]));
        b=b+1;
      %Case 1 and 4: More than 2 contiguous or discrete points
      else
        indx=zeros(length(temp),1);
        l=1;
        for j=2:length(temp) %contiguous points have been labelled sequen-
tially
          if (temp(j)-temp(j-1))==1
            indx(j-1)=l;
            indx(j)=l;
         elseif j>2 && (temp(j-1)-temp(j-2))==1 && (temp(j)-temp(j-1))~=1
            l=l+1;
          end
        end
        if max(indx)==0
          m=round(mean(close(temp)));
          typeMF(i)=1;
          UMF(i,:)=max(UMF(i,:),gaussmf(x,[sd(b) m]));
          LMF(i,:)=min(LMF(i,:),gaussmf(x,[sd(b) m]));
          b=b+1;
        else
%           c=0;
          for j=1:max(indx)
                temp1=temp(indx==j); % selecting days where the contigu-
ous points occur
            m=round(mean(close(temp1)));
%             if sd>=1
              UMF(i,:)=max(UMF(i,:),gaussmf(x,[sd(b) m]));
              LMF(i,:)=min(LMF(i,:),gaussmf(x,[sd(b) m]));
              b=b+1;
%               c=c+1;
%             end
          end
%           if c==1
%             typeMF(i)=1;
%           else
           typeMF(i)=2;
%           end
```

```
        end
      end
    end
    %Ensuring flat top
    loc1=0;loc2=0;
    for i=1:size(UMF,1)
      for j=1:1:size(UMF,2)
        if UMF(i,j)>0.999
          loc1=j;
          break;
        end
      end
      if j~=size(UMF,2)
        for j=size(UMF,2):-1:1
          if UMF(i,j)>0.999
            loc2=j;
            break;
          end
        end
      end
      if loc1~=0
        UMF(i,loc1:loc2)=1;
      end
    end
  end
  %%%%%%%%%%%%%%
  % FUZZY Secondary Factor Variation



  function BSF=fuzzyvarSF(VarSF)
  BSF=zeros(length(VarSF),1);
  for i=2:length(VarSF)
    for j=1:14
      if VarSF(i)<-6
        BSF(i)=1;
      elseif VarSF(i)>=6
        BSF(i)=14;
      elseif VarSF(i)>=(j-1)-6 && VarSF(i)<j-6
        BSF(i)=j+1;
      end
    end
  end
  end
```

```
%%%%%%%%%%%%%%%%%%%%%%%%%%
%% MF Using DE

function     [UMF,LMF,typeMF,rmse]=MFusingDE(A,close,x,sd,TestCP,TestA,
FLRG,WBS,TestB)
  genmax=50;
  F=0.2; % Scale Factor
  Cr=0.9; % Cross-over probability
  NP=20; % no. of population members
  gen=1;
  %% Initialization
  Zmin=ones(1,length(sd))*0.1;
  Zmax=sd;
  Z=zeros(NP,length(sd));
  for i=1:NP
    Z(i,:)=Zmin+rand*(Zmax-Zmin);
  end
  %%
  rmse=zeros(genmax,NP);
  while(gen<=genmax)
    disp('Gen=');
    disp(gen);
    %% Mutation
    V=zeros(NP,length(sd));
    for i=1:NP
      j=datasample(1:NP,1);
      while j==i
        j=datasample(1:NP,1);
      end
      k=datasample(1:NP,1);
      while k==i || k==j
        k=datasample(1:NP,1);
      end
      l=datasample(1:NP,1);
      while l==i || l==j || l==k
        l=datasample(1:NP,1);
      end
      V(i,:)=Z(j,:)+F.*(Z(k,:)-Z(l,:));
      for j=1:length(sd) % Ensuring V(i,j) is within Zmax and Zmin
        if V(i,j)<Zmin(j)
          V(i,j)=Zmin(j);
        elseif V(i,j)>Zmax(j)
          V(i,j)=Zmax(j);
        end
      end
```

```
    end
    %% Crossover
    U=zeros(NP,length(sd));
    for i=1:NP
      for j=1:length(sd)
        if rand<=Cr
          U(i,j)=V(i,j);
        else
          U(i,j)=Z(i,j);
        end
      end
    end
    %% Selection
    for i=1:NP
    [UMFu,LMFu,typeMFu]=formingMF(A,close,x,U(i,:));
    [UMFz,LMFz,typeMFz]=formingMF(A,close,x,Z(i,:));
      if evalfit(x,TestCP,TestA,UMFu,LMFu,typeMFu,FLRG,WBS...
           ,TestB)<evalfit(x,TestCP,TestA,UMFz,LMFz,typeMFz...
           ,FLRG,WBS,TestB)
        Z(i,:)=U(i,:);
      end
    end
    %% Storing fitness over generations
    for i=1:NP
      [UMFz,LMFz,typeMFz]=formingMF(A,close,x,Z(i,:));
      rmse(gen,i)=evalfit(x,TestCP,TestA,UMFz,LMFz,typeMFz...
         ,FLRG,WBS,TestB);
    end
    %%
    gen=gen+1;
  end
  [~,indx]=min(rmse(genmax,:));
  [UMF,LMF,typeMF]=formingMF(A,close,x,Z(indx,:));
End
%%%%%%%%%%%%%%%%%%%%%%%%%%%%%%
%% Over All Variation

function VarSF=overallvar(VarMF,SF,e,n)
  %
  VarSF1=zeros(size(SF,1),n);
  for i=1:n
    for j=2:size(SF,1)
      VarSF1(j,i)=(SF(j,i)-SF(j-1,i))*100/SF(j-1,i);
    end
  end
```

```
    %
    %
    tempDiffer=zeros(e,n);
    for i=1:n
      for j=3:e
        tempDiffer(j,i)=abs(VarSF1(j-1,i)-VarMF(j));
      end
    end
    DifferSF=sum(tempDiffer);
    %
    %
    WVSF=zeros(1,n);
    for i=1:n
      WVSF(i)=sum(DifferSF)/DifferSF(i);
    end
    %
    %
    WSF=zeros(1,n);
    for i=1:n
      WSF(i)=WVSF(i)/sum(WVSF);
    end
    %
    %
    VarSF=zeros(1,size(SF,1));
    for i=1:size(SF,1)
      VarSF(i)=sum(VarSF1(i,:).*WSF);
    end
    %
End
%%%%%%%%%%%%%%%%%%%%%%%%%
%% Partitioning The Universe of Discourse(UOD)



function [A,B,Var,Au,Al]=partitioning(CP)
%%
A=zeros(length(CP),1);
B=zeros(length(CP),1);
Var=zeros(length(CP),1);
%%
```

```
templ=min(CP);
tempu=max(CP);

if (roundn(templ,2)-templ)>0
  Al=roundn(templ,2)-100;
else
  Al=roundn(templ,2);
end

% if (roundn(tempu,2)-tempu)>0
%   Au=roundn(tempu,2);
% else
%   Au=roundn(tempu,2)+200;
% end
Au=Al;
while(1)
  Au=Au+200;
  if Au>tempu
    break;
  end
end

nFS=(Au-Al)/200;

for i=1:length(CP) %partioning the close series of main factor
  for j=1:nFS
    if CP(i)>=(j-1)*200+Al && CP(i)<j*200+Al
      A(i,1)=j;
      break;
    end
  end
end
%%
for i=2:length(CP) % finding the var series
    Var(i)=(CP(i)-CP(i-1))*100/CP(i-1);
end

for i=2:length(Var) % partioning the var series
  for j=1:14
    if Var(i)<-6
      B(i)=1;
    elseif Var(i)>=6
      B(i)=14;
    elseif Var(i)>=(j-1)-6 && Var(i)<j-6
      B(i)=j+1;
```

```
      end
    end
end
end
%%%%%%%%%%%%%%%%%%%%%
%% Plot RMSE with adaptation

function plotDE(rmse)
   figure,plot(1:size(rmse,1),min(rmse,[],2),'kx-','MarkerSize',5);
   xlabel('Generations -->');
   ylabel('RMSE -->');
   title('Evolving parameters to minimize RMSE');
end
%%%%%%%%%%%%%%%%%%%%%%%%%%
%% Plot Forecasted Price

function plotforecasted(predicted,TestCP,RMSE)
figure,
subplot(3,2,[1 2 3 4]);
plot(TestCP,'k*:');
hold on
plot(predicted,'ko-');
ylabel('Close Price');
axis([0 length(TestCP)+5 min(min(predicted),min(TestCP))-1000 max(max
(predicted),max(TestCP))+1000]);
hold off
legend('Actual','Predicted','location','SouthEast');
subplot(3,2,[5 6]);
stem(TestCP-predicted,'ko-');
text(27,max(TestCP-predicted)+50,{'RMSE=',RMSE});
axis([0  length(TestCP)  min(TestCP-predicted)  max(TestCP-predicted)
+150]);
xlabel('Testing days');
ylabel('Error');
end
%%%%%%%%%%%%%%%%%%%%%
%% Plot FOU(Foot Print of Uncertainity)

function plotFOU(UMF,LMF,x)
  for i=1:size(UMF,1)
     figure,shadedplot(x,LMF(i,:),UMF(i,:),[0.8 0.8 0.8]);
     xlabel('Close');
     ylabel('Membership values');
     title(['FOU for fuzzy set A', num2str(i)]);
  end
```

```
end
%%%%%%%%%%%%%%%%%%%%%%%%%%%%%%%
%% Plot Partitions

function plotpartitions(close,A,Al,Au)
  numFS=unique(A); % converting the close series into a matrix
  FS=zeros(length(numFS),length(close));
  for i=1:length(A)
    FS(A(i),i)=close(i);
  end

  % Plot Input Close
  plot(close,'k*-');
  hold on;
  for i=1:length(numFS)
    part=Al+(i-1)*200;
    plot([1 length(close)],[part part],'k:');
    hold on;
  end
  hold off;
  axis([1 length(close) Al Au]);
  xlabel('Training days');
  ylabel('Close');
  title('Fuzzifying training data');
  %
End
%%%%%%%%%%%%%%%%%%%%%%%
%% PredictionFunction

function forecasted=predict(TestCP,A,Al,Au,UMF,LMF,typeMF,FLRG,WBS,B
x=Al:Au-1;
AB=horzcat(A,B);
forecasted=zeros(length(TestCP),1);
for i=2:size(AB,1)
  a=AB(i-1,1);
  b=AB(i-1,2);
  consequent=find(FLRG(a,:,b));
  centroid=zeros(1,size(FLRG,1));
  if isempty(consequent)
    if typeMF(a)==1
      forecasted(i-1)=sum(x.*UMF(a,:))/sum(UMF(a,:));
    elseif typeMF(a)==2
      avgMF=(LMF(a,:)+UMF(a,:))/2;
      forecasted(i-1)=sum(x.*avgMF)/sum(avgMF);
%       forecasted(i-1)=KMmethod(LMF(a,:),UMF(a,:),x);
```

```
      end
    else
      %SF1=0;SF2=0;SF3=0;
      NF=0;
      for j=1:length(consequent)
        %Case1: a=T1FS, consequent(j)=T1FS
        if typeMF(a)==1 && typeMF(consequent(j))==1
          predy=interp1(x,UMF(a,:),TestCP(i-1),'linear','extrap');
          temp=ones(1,length(x))*predy;
          projMF=min(temp,UMF(consequent(j),:));
          centroid(consequent(j))=sum(x.*projMF)/sum(projMF);
        %Case2: a=T1FS, consequent(j)=IT2FS
        elseif typeMF(a)==1 && typeMF(consequent(j))==2
          predy=interp1(x,UMF(a,:),TestCP(i-1),'linear','extrap');
          temp=ones(1,length(x))*predy;
          projUMF=min(temp,UMF(consequent(j),:));
          projLMF=min(temp,LMF(consequent(j),:));
          avgMF=(projLMF+projUMF)/2;
          centroid(consequent(j))=sum(x.*avgMF)/sum(avgMF);
%           centroid(consequent(j))=KMmethod(projLMF,projUMF,x);
        %Case3: a=IT2FS, consequent(j)=T1FS
        elseif typeMF(a)==2 && typeMF(consequent(j))==1
          predU=interp1(x,UMF(a,:),TestCP(i-1),'linear','extrap');
%           predL=interp1(x,LMF(a,:),TestCP(i-1),'linear','extrap');
          temp=ones(1,length(x))*predU;
          projMF=min(temp,UMF(consequent(j),:));
          centroid(consequent(j))=sum(x.*projMF)/sum(projMF);
        %Case4: a=IT2FS, consequent(j)=IT2FS
        elseif typeMF(a)==2 && typeMF(consequent(j))==2
          predU=interp1(x,UMF(a,:),TestCP(i-1),'linear','extrap');
          predL=interp1(x,LMF(a,:),TestCP(i-1),'linear','extrap');
          tempU=ones(1,length(x))*predU;
          tempL=ones(1,length(x))*predL;
          projUMF=min(tempU,UMF(consequent(j),:));
          projLMF=min(tempL,LMF(consequent(j),:));
          avgMF=(projLMF+projUMF)/2;
          centroid(consequent(j))=sum(x.*avgMF)/sum(avgMF);
%           centroid(consequent(j))=KMmethod(projLMF,projUMF,x);
        end
        if j<a
            forecasted(i-1)=forecasted(i-1)+centroid(consequent(j))*FLRG
(a,consequent(j),b)*WBS(b,1);
        elseif j==a
            forecasted(i-1)=forecasted(i-1)+centroid(consequent(j))*FLRG
(a,consequent(j),b)*WBS(b,2);
```

```
        else
            forecasted(i-1)=forecasted(i-1)+centroid(consequent(j))*FLRG
(a,consequent(j),b)*WBS(b,3);
        end
      end
      for j=1:length(consequent)
        if j<a
          NF=NF+FLRG(a,consequent(j),b)*WBS(b,1);
        elseif j==a
          NF=NF+FLRG(a,consequent(j),b)*WBS(b,2);
        else
          NF=NF+FLRG(a,consequent(j),b)*WBS(b,3);
        end
      end
      forecasted(i-1)=forecasted(i-1)/NF;
  end
end
end
%%%%%%%%%%%%%%%%%%%%%%%%%%
%% Shade Plot

function [ha hb hc] = shadedplot(x, y1, y2, varargin)
y = [y1; (y2-y1)]';
ha = area(x, y);
set(ha(1), 'FaceColor', 'none') % this makes the bottom area invisible
set(ha, 'LineStyle', 'none')

% plot the line edges
hold on
hb = plot(x, y1, 'k', 'LineWidth', 2);
hc = plot(x, y2, 'k', 'LineWidth', 2);
hold off

% set the line and area colors if they are specified
switch length(varargin)
   case 0
   case 1
     set(ha(2), 'FaceColor', varargin{1})
   case 2
     set(ha(2), 'FaceColor', varargin{1})
     set(hb, 'Color', varargin{2})
     set(hc, 'Color', varargin{2})
   otherwise
end
```

```
% put the grid on top of the colored area
set(gca, 'Layer', 'top')
%%%%%%%%%%%%%%%%%%%%%%%%%%
%% Table IX

function FVG=tableIX(B_MF,B_SF)

  FVG=zeros(14,14);
  temp=zeros(14,14);

  for t=2:length(B_MF)
    Bx=B_MF(t);
    Bz=B_SF(t-1);
    temp(Bz,Bx)=1;
    FVG=FVG+temp;
    temp=zeros(14,14);
  end

end
%%%%%%%%%%%%%%%%%%%%%%
%% TableVI

function FLRG=tableVI(B,fromA,toA)

  endA=max([max(fromA),max(toA)]);
  FLRG=zeros(endA,endA,14);

  temp=zeros(endA,endA,14);
  for i=1:length(B)
    temp(fromA(i),toA(i),B(i))=1;
    FLRG=FLRG+temp;
    temp=zeros(endA,endA,14);
  end

end
%%%%%%%%%%%%%%%%%%%%%
%% Table X

function BS=tableX(FVG)
  BS=zeros(14,3);

  for i=1:14
    if sum(FVG(i,:))~=0
      BS(i,1)=sum(FVG(i,1:i-1))/sum(FVG(i,:));
```

```
        BS(i,2)=FVG(i,i)/sum(FVG(i,:));
        BS(i,3)=sum(FVG(i,i+1:14))/sum(FVG(i,:));
      end
   end
end
%%%%%%%%%%%%%%%%%%
```

How to run the program with workspace construction in MATLAB is available in the url: http://computationalintelligence.net/fuzzytimeseries/howtorun.html

## References


1. Engineering Statistic and book [Online]. http://www.itl.nist.gov/div898/handbook/pmc/section4/pmc41.htm
2. Chen, S. M., & Chen, C. D. (2011). Handling forecasting problems based on high-order fuzzy logical relationships. *Expert Systems with Applications, 38*(4), 3857–3864.
3. Chen, S. M., & Tanuwijaya, K. (2011). Fuzzy forecasting based on high-order fuzzy logical relationships and automatic clustering techniques. *Expert Systems with Applications, 38*(12), 15425–15437.
4. Wu, C. L., & Chau, K. W. (2013). Prediction of rainfall time series using modular soft computing methods. *Elsevier, Engineering Applications of Artificial Intelligence, 26,* 997–1007.
5. Wu, C. L., Chau, K. W., & Fan, C. (2010). Prediction of rainfall time series using modular artificial neural networks coupled with data-preprocessing techniques. *Elsevier, Journal of Hydrology, 389*(1–2), 146–167.
6. Barnea, O., Solow, A. R., & Stonea, L. (2006). On fitting a model to a population time series with missing values. *Israel Journal Of Ecology & Evolution, 52,* 1–10.
7. Chen, S. M., & Hwang, J. R. (2000). Temperature prediction using fuzzy time series. *IEEE Transactions on Systems, Man, and Cybernetics. Part B, Cybernetics, 30*(2), 263–275.
8. Song, Q., & Chissom, B. S. (1993). Fuzzy time series and its model. *Fuzzy Sets and Systems, 54*(3), 269–277.
9. Song, Q., & Chissom, B. S. (1993). Forecasting enrollments with fuzzy time series—Part I. *Fuzzy Sets and Systems, 54*(1), 1–9.
10. Song, Q., & Chissom, B. S. (1994). Forecasting enrollments with fuzzy time series—Part II. *Fuzzy Sets and Systems, 62*(1), 1–8.
11. Song, Q. (2003). A note on fuzzy time series model selection with sample autocorrelation functions. *Cybernetics &Systems, 34*(2), 93–107.
12. Jalil, A., & Idrees, M. (2013). Modeling the impact of education on the economic growth: Evidence from aggregated and disaggregated time series data of Pakistan. *Elsevier Economic Modelling, 31,* 383–388.
13. Edwards, R. P., Magee, J., & Bassetti, W. H. C. (2007). *Technical analysis of Stock Trends, 9th Edition, p. 10*. New York, USA: Amacom.
14. Raynor, W. J., Jr. (1999) *The international dictionary of artificial intelligence*. IL: Glenlake Publishing.
15. Mallat, S. (1997). *A wavelet tour of signal processing*. San diego, CA, USA: Academic Press.
16. Bhowmik, P., Das, S., & Konar, A., Nandi, D., & Chakraborty, A. (2010) Emotion clustering from stimulated encephalographic signals using a Duffing Oscillator. *International Journal of Computers in Healthcare, 1*(1).

17. Gupta, S. C., & Kapoor, V. K. (2002). *Fundamental of mathematical statistics*. New Delhi, India: S. Chand & Sons.
18. Jansen,B. H., Bourne, J. R., Ward, J. W. (1981). Autoregressive estimation of short segment spectra for computerized EEG analysis. *IEEE Transactions on Biomedical Engineering, 28*(9).
19. Hjorth, B. (1970). EEG analysis based on time domain properties. *Electroencephalography and Clinical Neurophysiology, 29,* 306–310.
20. Sanei, S. (2013) *Adaptive processing of brain signals*. USA: Wiley.
21. Chen, S. M. (1996). Forecasting enrollments based on fuzzy time series. *Fuzzy Sets and Systems, 81*(3), 311–319.
22. Hwang, J. R., Chen, S. M., & Lee, C. H. (1998). Handling forecasting problems using fuzzy time series. *Fuzzy Sets and Systems, 100*(1–3), 217–228.
23. Lee, L. W., Wang, L. H., & Chen, S. M. (2007). Temperature prediction and TAIFEX forecasting based on fuzzy logical relationships and genetic algorithms. *Expert Systems with Applications, 33*(3), 539–550.
24. Wong, W. K., Bai, E., Chu, A. W. C. (2010). Adaptive time-variant models for fuzzy-time-series forecasting. *IEEE Transaction on Systems, Man, Cybernetics-Part B: Cybernetics, 40*(6).
25. Cheng, C. H., Chen, T. L., Teoh, H. J., & Chiang, C. H. (2008). Fuzzy time-series based on adaptive expectation model for TAIEX forecasting. *Elsevier, Expert Systems with Applications, 34,* 1126–1132.
26. Chen, T. L., Cheng, C. H., & Teoh, H. J. (2007). Fuzzy time-series based on Fibonacci sequence for stock price forecasting. *Elsevier, Physica A, 380,* 377–390.
27. Huarng, K., & Yu, T. H. K. (2006). The application of neural networks to forecast fuzzy time series. *Physica A, 363*(2), 481–491.
28. Yu, T. H. K., & Huarng, K. H. (2010). A neural network-based fuzzy time series model to improve forecasting. *Expert Systems with Applications, 37*(4), 3366–3372.
29. Kuo, I. H., Horng, S. J., Kao, T. W., Lin, T. L., Lee, C. L., & Pan, Y. (2009). An improved method for forecasting enrollments based on fuzzy time series and particle swarm optimization. *Expert Systems with Applications, 36*(3), 6108–6117.
30. Lu, W., Yang, J., Liu, X., & Pedrycz, W. (2014). The modeling and prediction of time-series based on synergy of high order fuzzy cognitive map and fuzzy c-means clustering. *Elsevier, Knowledge Based Systems, 70,* 242–255.
31. Yu, T. H. K., & Huarng, K. H. (2008). A bivariate fuzzy time series model to forecast the TAIEX. *Expert Systems with Applications, 34*(4), 2945–2952.
32. Yu, T. H. K., & Huarng, K. H. (2010). Corrigendum to "A bivariate fuzzy time series model to forecast the TAIEX". *Expert Systems with Applications, 37*(7), 5529.
33. Chen, S. M., & Chang, Y. C. (2010). Multi-variable fuzzy forecasting based on fuzzy clustering and fuzzy rule interpolation techniques. *Information Sciences, 180*(24), 4772–4783.
34. Wong,H. L., Tu, Y. H., & Wang, C. C. (2009). An evaluation comparison between multivariate fuzzy time series with traditional time series model for forecasting Taiwan export. In *Proceedings of World Congress on Computer Science Engineering*, pp. 462–467.
35. Chen, S. M., & Tanuwijaya, K. (2011). Multivariate fuzzy forecasting based on fuzzy time series and automatic clustering clustering techniques. *Expert Systems with Applications, 38* (8), 10594–10605.
36. Huarng, K., Yu, H. K., & Hsu, Y. W. (2007). A multivariate heuristic model for fuzzy time-series forecasting. *IEEE Transactions on Systems, Man, and Cybernetics. Part B, Cybernetics, 37*(4), 836–846.
37. Sun, B. Q., Guo, H., Karimi, H. R., Ge, Y., Xiong, S. (2015). Prediction of stock index futures prices based on fuzzy sets and multivariate fuzzy time series. *Neurocomputing, Elsevier, 151,* 1528–1536.
38. Chen, T. L., Cheng, C. H., & Teoh, H. J. (2008). High-order fuzzy time-series based on multi-period adaptation model for forecasting stock markets. *Elsevier, Physica A, 387,* 876–888.

39. Zadeh, L. A. (1965). Fuzzy sets. *Information and Control, 8,* 338–353.
40. Zhang, W. (1992). *Applications engineer*, Aptronics Incorporated, Copyright © by Aptronix Inc.
41. Virkhareg, N., & Jasutkar, R. W. (2014) Neuro-fuzzy controller based washing machine. *International Journal of Engineering Science Invention, 3*, 48–51.
42. Konstantinidis, K., Gasteratos, A., & Andreadis, I. (2005). Image retrieval based on fuzzy color histogram processing. *Elsevier, Optics Communications, 248,* 375–386.
43. Halder, A., Konar, A., Mandal, R., Chakraborty, A., Bhowmik, P., Pal, N. R., et al., (2013) General and interval type-2 fuzzy face-space approach to emotion recognition. *IEEE Transactions On Systems, Man, And Cybernetics: Systems, 43*(3).
44. Karnik, N. N., & Mendel, J. M. (1999). Applications of type-2 fuzzy logic systems to forecasting of time-series. *Elsevier, Information Sciences, 120,* 89–111.
45. Huarng, K., & Yu, H. K. (2005). A type 2 fuzzy time series model for stock index forecasting. *Physica A, 353,* 445–462.
46. Bajestani, N. S., & Zare, A. (2011). Forecasting TAIEX using improved type 2 fuzzy time series. *Expert Systems with Applications, 38*(5), 5816–5821.
47. Chen, S. M., & Chen, C. D. (2011). TAIEX forecasting based on fuzzy time series and fuzzy variation groups. *IEEE Transactions on Fuzzy Systems, 19*(1), 1–12.
48. Mendel, J. M., & Wu, D. (2010). *Perceptual computing: Aiding people in making subjective judgements*. Hoboken, NJ: IEEE-Wiley press.
49. Chakraborty, S., Konar, A., Ralescu, A., & Pal, N. R. (2015). A fast algorithm to compute precise type-2 centroids for real time control applications. *IEEE Transaction on Cybernetics, 45*(2), 340–353.
50. Chen,S. M., Chu, H. P., Sheu, T. W. (2012). TAIEX forecasting using fuzzy time series and automatically generated weights of multiple factors. *IEEE Transaction on Systems, Man, and Cybernetics-Part A: Systems and Humans, 42*(6), 1485–1495.
51. Mendel,J. M., Hagras, H., Tan, W. W., Melek, W. W., & Ying, H. (2014) *Introduction to type-2 fuzzy logic control: Theory and applications*. USA: IEEE-Wiley Press.
52. Nilsson, N. J. (1980). *Principles ofartificial intelligence*. CA: Morgan Kaufmann.
53. Web Link: http://www.computationalintelligence.net/ieee_cyb/index_prediction_appendix1.pdf
54. Garibaldi,J. M. (2010). Alternative forms of non-standard fuzzy sets: A discussion chapter. In *Proceedings of Type-2 Fuzzy Logic: State of the Art and Future Directions*, London.
55. Wu, D. (2011). A constraint representation theorem for interval type-2 fuzzy sets using convex and normal embedded type-1 fuzzy sets, and its application to centroid computation. In *Proceedings of World Conference on Soft Computing*, San Francisco, CA.
56. Web Link for historical Data TAIEX. [Online]. Available: http://www.twse.com.tw/en/products/indices/tsec/taiex.php
57. Storn, R., & Price, K. G. (1997). Differential Evolution- a simple and efficient heuristic for global optimization over continuous spaces. *Journal of Global Optimization, 11*(4), 341–359.
58. Das, S., & Suganthan, P. N. (2011) Differential evolution: A survey of the state-of-the-art. *IEEE Transaction on Evolutionary Computation, 15*(1), 4–31.
59. Chen, S. M., Kao, P.-Y. (2013). TAIEX forecasting based on time series, particle swarm optimization techniques and support vector machines. *Information Science, Elsevier, 247,* 62–71.
60. Cai, Q., Zhang, D., Zheng, W., & Leung, S. C. H. (2015). A new fuzzy time series forecasting model combined with ant colony optimization and auto-regression. *Knowledge-Based Systems, 74,* 61–68.
61. Chen, Mu-Yen. (2014). A high-order fuzzy time series forecasting model for internet stock trading. *Future Generation Computer Systems Elsevier, 37,* 461–467.
62. Gardner, E., & McKenzie, E. (1989). Seasonal exponential smoothing with damped trends. *Management Science, 35*(3), 372–376.

63. Kuremoto, T., Kimura, S., Kobayashi, K., & Obayashi, M. (2014). Time series forecasting using a deep beliefnetwork with restricted boltzmann machines (online open access). *Neurocomputing, 137*(5), 47–56.
64. Yildiz,O. T. (2013) Omnivariate rule induction using a novel pairwise statistical test. *IEEE Transactions On Knowledge And Data Engineering*, *25*(9).

# Chapter 3
# Handling Main and Secondary Factors in the Antecedent for Type-2 Fuzzy Stock Prediction

**Abstract** Traditional fuzzy logic based approaches to prediction of stock index time-series utilize the reasoning mechanisms of type-1 fuzzy sets. The predictions undertaken thereby occasionally suffer from representational uncertainty. This chapter introduces interval type-2 fuzzy reasoning to capture the uncertainty buried under the individual partitions of a time-series. It presents three different methods of autonomous construction of membership functions, and one additional method for automatic adaptation of membership function for further tuning of memberships with the latest data of the time-series. The first method employs interval type-2 fuzzy reasoning to predict the next day variation in main factor time-series from its current value. The second method too introduces an interval type-2 reasoning with secondary factor variation as an additional antecedent for the prediction. It organizes the dynamic range of the (main factor) time-series as non-uniformly partitioned segments using evolutionary algorithm, so that each partition includes at least one data point sufficient to capture the uncertainty by interval type-2 model. The third method employs uniform partitioning with no restriction on the number of data points in the partitions. It employs type-1 fuzzy sets to capture the uncertainty in a partition when it includes a single block of contiguous data and an interval type-2 fuzzy set when the partition includes two or more blocks. The last method involves additional tuning of the membership functions with recent data from the time-series to imbibe the prediction results with the current trends. Experiments undertaken reveal that the third method with provisions for adaptation of membership functions with recent data outperforms the first two methods. The said method also outperforms existing techniques by a large margin of root mean square error.

## 3.1 Introduction

A time-series represents discrete samples of a time-valued function obtained at uniform intervals of time [1]. Prediction of the next sample value of a time-series from its current sample values is generally regarded as the time-series prediction



A. Konar and D. Bhattacharya, *Time-Series Prediction and Applications*,
Intelligent Systems Reference Library 127,
DOI 10.1007/978-3-319-54597-4_3

problem. In an economic time-series, the next sample value is presumed to depend only on the current sample value [2, 3] and other external factors [4–6] where the latter in many circumstances are not completely known. This chapter attempts to predict an economic time-series considering other standard time-series as influencing factors in the prediction process. The other standard time-series that may influence the prediction of given time-series is called secondary factor.

Quite a few interesting chapters on time-series prediction are available in the literature [7–14]. A few selected chapters that need special mention include neural network models [15–17], regression analysis [18], fuzzy techniques [7, 10–13, 19] evolutionary algorithms [20], adaptation models [21–23] automatic clustering [3, 24, 25], heuristic model [26], genetic algorithm [19], and Fibonacci sequence [27]. In the present work, we employ fuzzy models for their inherent potential to handle uncertainty in the prediction process. The first successful work employing fuzzy models for economic time-series prediction [12] is due to Song et al. [10–13]. In [10, 11], the authors employed fuzzy relational algebra for time-series prediction. Several extensions [4, 5, 8, 10, 14, 16, 17, 19–34] to the primitive fuzzy model introduced by Song et al. have been developed in the past to minimize the root mean square error of time-series prediction. However, unfortunately, none of the existing models could capture the inherent uncertainty hidden in the time series or its secondary factors. This chapter attempts to employ Interval type-2 [35] fuzzy sets (IT2FS) to model both intra- and inter-personal level uncertainty in a time-series.

Most of the fuzzy logic based time-series prediction algorithms use type-1 [39] fuzzy reasoning for prediction. Very recently, researchers took keen interest to examine the role of interval type-2 fuzzy sets (IT2FS) [35–38, 40] for time-series prediction. Unfortunately, none of the above prediction algorithms used traditional IT2FS in its true spirit, as indicated by Mendel and his research group [41]. For instance, in [37, 38], the authors employ three type-1 fuzzy sets HIGH, LOW and CLOSE and consider their union together to represent an IT2FS. Unfortunately, the type-1 fuzzy sets used to develop the IT2FS should be of same genre [40–42] i.e., they all should characterize the same fuzzy concept such as CLOSE. This chapter, however, considers type-1 fuzzy sets of the same genre to construct the IT2FS. In addition, propositional logic [37, 38] instead of traditional fuzzy logic, is used to perform fuzzy reasoning, which may not be acceptable to practitioners of fuzzy sets and logic. The present work strictly follows the formalisms of IT2FS [43, 44] to develop the algorithm for time-series prediction.

Although there exist traces of works on using secondary indices [4, 6, 9] for time-series prediction, we feel that an alternative but more simplified method to address the same problem can be developed by embedding the last sample values of both main factor time-series and secondary factor in the premise of the fuzzy production rules used for prediction. In fact, the present work employs such production rule to infer the current value of the main factor time-series. The objective of the present work thus is two-fold. First, we attempt to design a general framework of reasoning with rules indicated above, where the type of fuzzy propositions present in the premise and consequent are determined by the time-series data only and not by the user. For example, if a partition of the time series contains one or a few contiguous data point, the membership function selected to represent those fragments of data is type-1. On the other hand, if a partition includes several contiguous blocks of data points, an IT2FS representation is a better choice for the data points in the partition.

Secondly, because of allowance of fuzzy propositions in premise and/or consequent as type-1 or IT2FS, the reasoning mechanism may take eight different forms with either or both propositions in premise and the single proposition in consequent to take either of two forms. Naturally, for each of the eight cases, we need to develop a method of fuzzy reasoning. Lastly, to get back the predicted results in real quantity, we need to use a process of de-fuzzification to transform a fuzzy inference back to real data. Two possible modalities of de-fuzzification apply in the present context, depending on the type of the inference. If it is a type-1 fuzzy set, we use type-1 defuzzification. Otherwise, we go for IT2 defuzzification using the well-known Karnik-Mendel algorithm [41].

Finally, we consider a system validation approach to tune membership functions with an ultimate aim to reduce the root mean square prediction error. This is done using an evolutionary approach realized with Differential Evolution algorithm. Any traditional evolutionary algorithm could be used to solve the problem. However, we use Differential Evolution [52] for pour past experience of using it [49] along with its added benefit of low computational cost, few control parameters and its small length of code.

The rest of the chapter is fragmented into five sections. Section 3.2 provides preliminaries to IT2FS. In Sect. 3.3, we develop the algorithm for time-series prediction. Experimental results undertaken through computer simulation are given in Sect. 3.4. Conclusions are listed in Sect. 3.5.

## 3.2 Preliminaries

This section introduces a few definitions covering both time-series prediction and IT2FS. These definitional are used in the rest of the chapter.

**Definition 3.2.1** The ***close price*** $c(t)$ [45] of a stock index refers to the last traded price on the trading day t in a given stock exchange.

**Definition 3.2.2** The stock index under consideration for prediction of a time series is called *Main Factor Time Series* (MFTS). Here, we consider TAIEX (Taiwan Stock Exchange Index) as the MFTS [46].

**Definition 3.2.3** The indices associated with MFTS that have large influence on the prediction of the MFTS are called *Secondary Factor Time Series* (SFTS). Here, we consider NASDAQ (National Association of Securities and Dealers Automated Quotations) as the SFTS.

**Definition 3.2.4** The *Variation Time Series* (VTS) for a given close price time series (CTS) $c\,(t)$, is evaluated by,

$$VTS = c(t) - c(t-1) \tag{3.1}$$

for $t \in [t_{\min}, t_{\max}]$, where $t_{min}$ and $t_{max}$ denote the beginning and terminating days of the *training period* [4]. We here onwards use $V_M(t)$ and $V_S(t)$ to denote the VTS for MFTS (or MFVTS) and SFTS (or SFVTS) respectively.

**Definition 3.2.5** Prediction of an MFTS at time $t+1$ refers to evaluating $c(t+1)$ from its past Main factor VTS (also known as MFVS) values: $V_M(t)$, $V_M(t-1)$, …, $V_M(t-(m-1))$, obtained from MFTS values $c(t)$, $c(t-1)$, $c(t-2)$, $c(t-3)$, …, $c(t-(m-1))$ and secondary factor VTS (SFVTS also known as SFVS) $V_S(t), V_S(t-1), \ldots, V_S(t-(m-1))$ for some positive integer m. This is referred to as *m*th order forecasting. However, in most time-series forecasting [2, 4, 5, 9–13] including the one in the present chapter, 1st order forecasting model with $m = 1$ is used [6].

**Definition 3.2.6** A T1 fuzzy set is represented by a two tuple$\langle \mathrm{x}, \mu_{\mathrm{A}}(\mathrm{x})\rangle$, where x denotes a linguistic variable in a Universe of discourse X and $\mu_{\mathrm{A}}(\mathrm{x})$ denotes the membership value of x in the fuzzy set A, lying in $[0, 1]$.

**Definition 3.2.7** A general type-2 Fuzzy Set (GT2FS) [36] is a three tuple given by $\langle \mathrm{x}, \mu_{\mathrm{A}}(\mathrm{x}), \mu_{\mathrm{A}}(\mathrm{x}, \mu_{\mathrm{A}}(\mathrm{x}))\rangle$ where $x$ and $\mu_{\mathrm{A}}(\mathrm{x})$ have the same meaning as in Definition 3.2.6, and $\mu_{\mathrm{A}}(\mathrm{x}, \mu_{\mathrm{A}}(\mathrm{x}))$ is the secondary membership in [0, 1] for a given $(x, \mu_{\mathrm{A}}(\mathrm{x}))$.

**Definition 3.2.8** An interval type-2 fuzzy set (IT2FS) [35], [40, 42] is defined by two T1 membership functions (MFs), called *Upper Membership Function (UMF)*, and *Lower Membership Function (LMF)*. An IT2FS $\widetilde{A}$, therefore, is represented by $\langle \underline{\mu}_{\widetilde{A}}(x), \overline{\mu}_{\widetilde{A}}(x)\rangle$ where $\underline{\mu}_{\widetilde{A}}(x)$ and $\overline{\mu}_{\widetilde{A}}(x)$ denote the lower and upper membership functions respectively. The secondary membership $\mu_{\mathrm{A}}(\mathrm{x}, \mu_{\mathrm{A}}(\mathrm{x}))$ in IT2FS is considered as 1 for all $x$ and $\mu_{\mathrm{A}}(\mathrm{x})$.

**Definition 3.2.9** The left end point centroid is the smallest of all possible centroids of the embedded fuzzy sets [42–44] in an IT2FS $\widetilde{A}$ and is evaluated by

$$c_l = \frac{\sum_{i=1}^{k-1} \overline{\mu}_{\widetilde{A}}(x_i).x_i + \sum_{i=k+1}^{N} \underline{\mu}_{\widetilde{A}}(x_i).x_i}{\sum_{i=1}^{k-1} \overline{\mu}_{\widetilde{A}}(x_i) + \sum\limits_{i=k+1}^{N} \underline{\mu}_{\widetilde{A}}(x_i)} \tag{3.2}$$

using the well-known Karnik-Mendel algorithm [42–44] where $x \in \{x_1, x_2, \ldots, x_N\}$ and $x_{i+1} > x_i\ \forall i = 1$ to $N-1$. Here $x = x_k$ is a switch point and $N$ denotes the number of sample points of $\overline{\mu}_{\widetilde{A}}(x_i)$ and $\underline{\mu}_{\widetilde{A}}(x_i)$.

**Definition 3.2.10** The right end point centroid is the largest of all possible centroids of the embedded fuzzy sets [48, 49, 51] in an IT2FS $\widetilde{A}$ and is evaluated by

$$c_r = \frac{\sum_{i=1}^{k-1} \underline{\mu}_{\tilde{A}}(x_i).x_i + \sum_{i=k+1}^{N} \overline{\mu}_{\tilde{A}}(x_i).x_i}{\sum_{i=1}^{k-1} \underline{\mu}_{\tilde{A}}(x_i) + \sum_{i=k+1}^{N} \overline{\mu}_{\tilde{A}}(x_i)} \tag{3.3}$$

using the well-known Karnik-Mendel algorithm [42–44] where $x \in \{x_1, x_2, .., x_N\}$ and $x_{i+1} > x_i \; \forall i = 1$ to $N - 1$. Here $x = x_k$ is a switch point and $N$ denotes the number of sample points of $\overline{\mu}_{\tilde{A}}(x_i)$ and $\underline{\mu}_{\tilde{A}}(x_i)$.

**Definition 3.2.11** The centroid of an IT2FS is given by

$$c = \frac{(c_l + c_r)}{2} \tag{3.4}$$

where $c_l$ and $c_r$ are the left and the right end point centroids.

**Definition 3.2.12** The RMSE [4] for predicted values is defined as

$$RMSE = \sqrt{\frac{\sum_{i=1}^{n} (AV - FV)^2}{n}} \tag{3.5}$$

where n is the number of days predicted, AV is actual value, FV is forecasted value.

## 3.3 Proposed Approach

Traditional approaches to economic time-series prediction usually adopt uniform partitioning of the dynamic range of the series for its efficient prediction [2, 4–6, 9–13]. Choice of the number of partitions here plays a vital role on the accuracy of the prediction. The prediction accuracy seems to be increased for increasing number of partitions, however, with a risk of having one or fewer data points in a partition. This chapter offers two alternative approaches to handle the above problem. In Method-I and -II presented below, we consider non-uniform partitioning of the time-series, so as to ensure that each partition includes two or more disjoint blocks of contiguous data points with at least two consecutive data points in a block. The above restriction on the width of individual partition helps us develop a type-1 fuzzy model for each block of data in a partition, and an IT2FS model for a complete partition. In Method-III, we allow uniform-width partitioning, even with a single block (in a partition) containing at least one data point. In case there exists a single block in a partition, we represent the partition by a type-1 fuzzy set. However, partitions with two or more blocks are represented by IT2FS. Thus, method-III employs a mixed model of fuzzy reasoning, where antecedent and consequents of implication rule may include both type-1 and IT2 MFs.

The fundamental difference between Method-I and -II lies in the fact that method-I considers implication from MFVTS at day $(t - 1)$ to MFVTS at day $t$, whereas the method-II considers implication from MFVTS and SFVTS both at day $(t-1)$ to MFVTS at day $t$. The method-III is similar with method-II from the point of view of implication rules, but is different with respect to the types of partitioning. Lastly, even after doing all as stated above, we noticed that the results of prediction are sensitive to the choice of T1 MFs. This motivated us to develop method IV, where the T1 MFs are adapted using an evolutionary algorithm with an aim to reduce root mean square error (RMSE) [4].

### 3.3.1 Method-I: Prediction Using Classical IT2FS

**Step 1**: For a given time-series $c(t)$, we compute the main factor variation time-series as shown in Fig. 3.1 $v(\mathrm{t}) = c(t) - c(t-1)$ for all $t$.

**Step 2**: Partition the main factor variation time-series MFVS $v(t)$ into unequal sized $p$ intervals, called $P_i$ for $i = 1\ to\ p$, where the length of the $i$th interval is given $d_i$, such that

$$J = \left((v_{max} - v_{min}) - \sum_{i=1}^{n} d_i\right)^2 \tag{3.6}$$

is minimized, where $v^{max}$ and $v^{min}$ respectively denote the maximum and minimum values of the time-series respectively for a given finite range of time. This minimization problem has to satisfy two constraints. First, the number of blocks within any partition should be $\geq 2$. Second, the number of data points in a block should

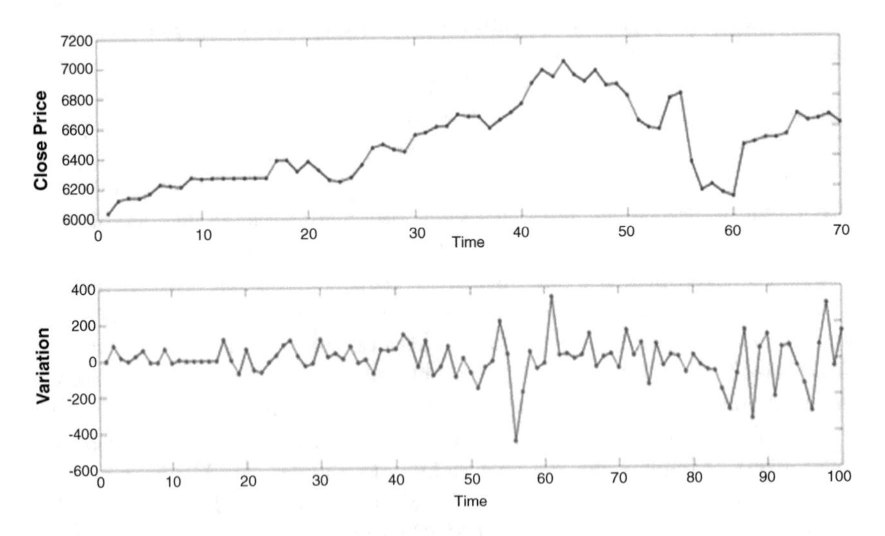


**Fig. 3.1** Close price time series and variation time series

$\geq 2$. This is realized by formulating an optimization problem with $d_i$, $for\ i = 1\ to\ n$ as unknown such that

$$J = \left( (v_{max} - v_{min}) - \sum_{i=1}^{n} d_i \right)^2 + w_1 \cdot Penalty_1 + w_2.Penalty_2$$

is minimized where

$$\begin{aligned} Penalty_1 &= 100, \text{ if number of blocks of contiguous data points } < 2, \\ &= 0, \text{ otherwise.} \end{aligned}$$

and

$$\begin{aligned} Penalty_2 &= 100, \text{ if number Penalty of data points within a block} < 2, \\ &= 0, \text{ otherwise.} \end{aligned}$$

Any evolutionary algorithm with a trial solution having components equal to the partition widths $d_1, d_2, \ldots, d_n$ can be used to solve the minimization problem. We here use the well-known Differential Evolution (DE) algorithm for its fast convergence, better accuracy and fewer control parameters and above all our familiarity with the algorithm for quite a long time [49, 51, 52]. The pseudo code of DE algorithm is given in the appendix.

**Step 3**: For each set of contiguous data points in a partition $P_i$, we construct a T1 triangular MF $A_{i,j}$, where the centre and base width of the isosceles triangular MF are equal to the respective mean and twice the variance of the data points within the contiguous data points and $j$ denotes a particular (here $j$th) type-1 membership function in a partition. In case there is a single isolated data point, we consider the centre of the triangular MF equal to the value of the actual data point and consider a small fixed base width of 0.001 units.

**Step 4**: For all T1 MFs $A_{i,j}$ lying in a partition $P_i$, we construct a single IT2 MF $\widetilde{A}_i$ whose UMF is obtained by taking the maximum of the MFs of all T1 FS within the partition. Similarly, the LMF of the IT2Fs is obtained by taking the minimum of the MFs of all the T1 fuzzy sets in a partition. Symbolically,

$$LMF(\widetilde{A}_i) = \underline{\mu}_{\widetilde{A}_i}(x) = \operatorname*{Min}_{j=1}^{j_{\max}}(\mu_{A_{i,j}}(x)) \tag{3.7}$$

and

$$UMF(\widetilde{A}_i) = \overline{\mu}_{\widetilde{A}_i}(x) = \operatorname*{Max}_{j=1}^{j_{\max}}(\mu_{A_{i,j}}(x)). \tag{3.8}$$

**Step 5**: To satisfy convexity criterion, the UMF is approximated by a flat-top representation. This is done by joining the first and last maxima of the UMF by a straight line parallel to the x-axis.

**Step 6**: For each pair of current day $t-1$ and next day $t$, determine the respective IT2FS $\widetilde{A}_i$ and $\widetilde{A}_j$, lying in partitions $P_i$ and $P_j$, and save the fuzzy relational mappings: $\widetilde{A}_i \rightarrow \widetilde{A}_j$ in a list.

**Step 7**: For a given day $t-1$, check the occurrence of $\widetilde{A}_i$ in the list. If $\widetilde{A}_i$ occurs in r number of rules, fire each rule with current value of $v(t-1)$ and generate IT2 inference using the following policy.

Let $(t-1)$th day's variation lies on $i$th partition. On firing the rules with antecedent $\widetilde{A}_i$:$\widetilde{A}_i \rightarrow \widetilde{A}_j, \widetilde{A}_i \rightarrow \widetilde{A}_k$, and $\widetilde{A}_i \rightarrow \widetilde{A}_l$ we determine the fuzzy inferences by the following procedure. Let UMF and LMF for $\widetilde{A}_i$ be $UMF_i$ and $LMF_i$ respectively. For the rule: $\widetilde{A}_i \rightarrow \widetilde{A}_j$, the IT2 inference is obtained by

$$UMF'_j = \mathrm{Min}[UMF_i(v'), UMF_j], \tag{3.9}$$

$$LMF'_j = \mathrm{Min}\,[LMF_i(v'), LMF_j], \tag{3.10}$$

where,$v'$ is a measures value of $v\,(t-1) \exists t$. Similarly, we obtain $UMF'_y$ and $LMF'_y$ by replacing index $j$ by $y$ for $y \in \{k, l\}$ in the remaining rules.

**Step 8**: Defuzzify each inference using (3.2), (3.3) and (3.4) and determine the final variation by taking weighted sum of each centroid by the probability of occurrence of the rule, i.e., $\bar{c} = \sum_j (w_{ij} \times c_j)$, where $c_j$ is the centroid of the inference obtained by consulting the rule $\widetilde{A}_i \rightarrow \widetilde{A}_j$.

The probability of occurrence for the rule $\widetilde{A}_i \rightarrow \widetilde{A}_j$ is $w_{ij} = \frac{f_j(A_i \rightarrow A_j)}{\sum_{\forall l} f_l(A_i \rightarrow A_l)}$, where $f_j$ and $f_l$ the frequency count of the rule $\widetilde{A}_i \rightarrow \widetilde{A}_j$ and $\widetilde{A}_i \rightarrow \widetilde{A}_l$ respectively.

**Step 9**: Next Predicted price at day $t$ i.e., c(t) is evaluated by the following step:

$$c(t) = c(t-1) + v(t),$$

where $\mathrm{v(t)} = \bar{c}$ the defuzzified signal amplitude, representing the predicted variation for the next day t.

### 3.3.2 *Method-II: Secondary Factor Induced IT2 Approach*

**Step 1**: Repeat step 1 of Method-I to compute the variation time-series $v(t) = c(t) - c(t-1)$ for all $t$, where $c(t)$ is the close price at day $t$.

**Step 2**: Partition the main factor variation time-series MFVS $v(t)$ into unequal sized $p$ intervals, called $P_i$ for $i = 1$ *to* $p$, following step 2 of Method-I.

**Step 3**: For each set of contiguous data points in a partition $P_i$, construct a T1 triangular MF $A_{i,j}$, where $j$ denotes a particular (here $j$th) type-1 membership function in a partition using step 3 of Method-I.

**Step 4**: For all T1 MFs $A_{i,j}$ lying in a partition $P_i$, we construct a single IT2 MF $\widetilde{A}_i$ whose UMF and LMF are obtained as

$$LMF\left(\widetilde{A}_i\right) = \underline{\mu}_{\widetilde{A}_i}(x) = \mathop{\mathrm{Min}}_{j=1}^{j_{\max}}(\mu_{A_{i,j}}(x)) \tag{3.11}$$

$$UMF(\widetilde{A}_i) = \overline{\mu}_{\widetilde{A}_i}(x) = \mathop{\mathrm{Max}}_{j=1}^{j_{\max}}(\mu_{A_{i,j}}(x)). \tag{3.12}$$

following step 4 of Method-I.

**Step 5**: Carry out flat-top approximation of the resulting IT2FS by joining the first and last maxima of the UMF by a straight line parallel to the x-axis, as indicated in step 5 of Method-I.

**Step 6**: For a given SFTS $c_s(t)$, compute the secondary factor variation series (SFVS) $v_s(t) = c_s(t) - c_s(t-1)$ for all $t$.

**Step 7**: Partition for the sake of simplicity the SFVS $v_S(t)$ into equal sized $q$ intervals, called $Q_j$, $j = 1$ *to* $q$, where the length of the intervals is given by $(v_S^{max} - v_S^{min})/q$, where $v_S^{max}$ and $v_S^{min}$ respectively denote the maximum and minimum values of the SFTS respectively for a given finite range of time.

**Step 8**: Construct a T1 Triangular MF $B_{i,j}$ for a set of contiguous data points in partition $Q_i$ following the same policy as used at step 3 of Method-I.

**Step 9**: Construct IT2MF $\widetilde{B}_i$ for all T1 MF $B_{i,j}$ lying on partition $Q_i$ following the same policy used in method 1 at step 4 and step 5.

**Step 10**: Determine the pair of IT2FS for MFVTS $\widetilde{A}_i$ and SFVTS $\widetilde{B}_j$ at current day $t-1$, lying in partitions $P_i$ and $Q_i$ respectively and also determine the IT2FS for MFVTS $\widetilde{A}_k$ at next day $t$, and save the fuzzy relational mapping: $\widetilde{A}_i, \widetilde{B}_j \rightarrow \widetilde{A}_k$ in a list $L$.

**Step 11**: For a given day $t-1$, check the occurrence of $\widetilde{A}_i$, $\widetilde{B}_j$ pair in all sets of rules of the form given above in the list L. If $\widetilde{A}_i$, $\widetilde{B}_j$ occur in $r$ number of rules, fire each rule with current value of $v_M(t-1)$, $v_S(t-1)$ and generate IT2 inference using the following policy.

(a) Let $(t-1)$th day's MFVTS lies on $i$th partition and SFVTS lies on $j$th partition of their respective universe of discourses. Let the rules found in the list L with antecedent $\widetilde{A}_i, \widetilde{B}_j$ include: $\widetilde{A}_i, \widetilde{B}_j \rightarrow \widetilde{A}_k; \widetilde{A}_i, \widetilde{B}_j \rightarrow \widetilde{A}_l; \widetilde{A}_i, \widetilde{B}_j \rightarrow \widetilde{A}_m$. Let UMF and LMF for $\widetilde{A}_i$ be $UMF_i$ and $LMF_i$, that for $\widetilde{B}_j$ be $UMF_j$ and $LMF_j$ respectively. For the rule: $\widetilde{A}_i, \widetilde{B}_j \rightarrow \widetilde{A}_k$ the IT2 inference is obtained by

$$umf = \text{Min}[UMF_i(v'_M), UMF_j(v'_S)] \tag{3.13}$$

$$lmf = \text{Min}[LMF_i(v'_M), LMF_j(v'_S)] \tag{3.14}$$

$$UMF'_k = \text{Min}[umf, UMF_k] \tag{3.15}$$

$$LMF'_k = \text{Min}[lmf, LMF_k] \tag{3.16}$$

where $v'_M$ is a measured value of $v_M(t-1) \exists t$. and $v'_S$ is a measured value of $v_S(t-1) \exists t$.

Similarly, we obtain $UMF'_y$ and $LMF'_y$ by replacing index $k$ by $y$ where $y \in \{l, m\}$ for the remaining rules.

**Step 12**: Defuzzification is done and forecasted value $\overline{c}$ is computed by the following step:

$\overline{c} = \sum_k (w_{i,j,k} \times c_k)$, where $c_k$ is the centroid of the inference obtained by consulting the rule $\widetilde{A}_i, \widetilde{B}_j \rightarrow \widetilde{A}_k$, where $w_{i,j,k}$ is obtained by

$$w_{i,j,k} = \frac{f_k(\widetilde{A}_i, \widetilde{B}_j \rightarrow \widetilde{A}_k)}{\sum_{\forall l} f_l(\widetilde{A}_i, \widetilde{B}_j \rightarrow \widetilde{A}_l)} \tag{3.17}$$

where $f_k$ is the frequency count of the rule $\widetilde{A}_i, \widetilde{B}_j \rightarrow \widetilde{A}_k$.

**Step 13**: The Predicted price at day $t$ i.e., c(t) is evaluated by the following step:

$$c\,(t) = c(t-1) + v(t),$$

where v(t) $= \overline{c}$ the defuzzified signal amplitude, representing the predicted variation for the next day t.

### 3.3.3 *Method-III: Prediction in Absence of Sufficient Data Points*

In Method-III, we partition a time-series into equal sized intervals for both MFVS and SFVS. Here, the partitions containing a single time block are represented by type-1 fuzzy sets, where the partitions containing two or more blocks are represented by IT2FS. The prediction algorithm introduced here uses type-1, IT2 and mixed (type1-IT2) reasoning depending on the types of the fuzzy sets used in the antecedents and consequents of the fired rules. Here, the antecedent part of the rules include MFVTS at day $(t-1)$ and SFVTS at day $(t-1)$, where the consequent part includes only MFVTS at day $t$. Naturally, considering each proposition in antecedent/consequent to be type-1 or IT2FS, we have eight possible reasoning.

**Algorithm for Time-Series Prediction Using Mixed Type-1 and IT2 Fuzzy Sets in the Rule**

**Step 1**: Repeat Step 1 of Method-I to compute the variation time-series $v(t) = c(t) - c(t-1)$ for all $t$, where $c(t)$ is the close price at day $t$.

**Step 2**: Partition the main factor variation time-series MFVS $v(t)$ into equal sized $p$ intervals, called $P_i$ for $i = 1 \; to \; p$, where the length of the intervals is given by $(v_2 - v_1)/p$, where $v_2 = v^{max} + D_2$ and $v_1 = v^{\min} - D_1$, here $D_1$ and $D_2$ are two positive real numbers such that $(v_2 - v_1)/p$ is an integer number, where $D_1, D_2 < (v_2 - v_1)/p$, $v^{max}, v^{\min}$ respectively denote the maximum and minimum values of the time-series respectively for a given finite range of time.

**Step 3**: For each set of contiguous data points in a partition $P_i$, construct a T1 triangular MF $A_{i,j}$, where $j$ denotes a particular (here $j$th) type-1 membership function in a partition using step 3 of Method-I.

**Step 4**: *For* T1 MFs $A_{i,j}$ lying in a partition $P_i$,

$$IF \; j \in [1, j_{max}] \quad and \quad j_{max} > 1$$

we construct a single IT2 MF $\widetilde{A}_i$ whose LMF and UMF are obtained as

$$LMF(\widetilde{A}_i) = \underline{\mu}_{\widetilde{A}_i}(x) = \underset{j=1}{\overset{j_{max}}{\mathrm{Min}}}(\mu_{A_{i,j}}(x)) \tag{3.18}$$

$$UMF(\widetilde{A}_i) = \overline{\mu}_{\widetilde{A}_i}(x) = \underset{j=1}{\overset{j_{max}}{\mathrm{Max}}}(\mu_{A_{i,j}}(x)). \tag{3.19}$$

following step 4 of Method-I,

*Else*$(j_{max} = 1)$ consider $A_i = A_{i,1}$ as a T1 MF.

*End—For;*

**Step 5**: Carry out flat-top approximation of the resulting IT2FS $\widetilde{A}_i$ by joining the first and last maxima of the UMF by a straight line parallel to the x-axis, as indicated in Step 5 of Method-I.

**Step 6**: For a given SFTS $c_s(t)$, compute the secondary factor variation series (SFVTS) $v_s(t) = c_s(t) - c_s(t-1)$ for all $t$.

**Step 7**: Partition the SFVTS $v_S(t)$ into equal sized $q$ intervals, called $Q_j, j = 1 \; to \; q$, where the length of the intervals is given by $(v_S^{max} - v_S^{min})/q$, where $v_S^{max}$ and $v_S^{min}$ respectively denote the maximum and minimum values of the SFVTS respectively for a given finite range of time.

**Step 8**: Construct a T1 Triangular MF $B_{i,j}$ for a set of contiguous data points in partition $Q_i$ following the same policy as used at Step 3 of Method-I.

**Step 9**: *For* T1 MFs $B_{i,j}$ lying in a partition $Q_i$,

$$IF\ j \in [1, j_{max}] \quad and \quad j_{max} > 1$$

we construct a single IT2 MF $\widetilde{B}_i$ whose UMF and LMF are obtained as

$$LMF(\widetilde{B}_i) = \underline{\mu}_{\widetilde{B}_i}(y) = \underset{j=1}{\overset{j_{\max}}{\mathrm{Min}}}(\mu_{B_{i,j}}(y)) \tag{3.20}$$

$$UMF(\widetilde{B}_i) = \overline{\mu}_{\widetilde{B}_i}(y) = \underset{j=1}{\overset{j_{\max}}{\mathrm{Max}}}(\mu_{B_{i,j}}(y)). \tag{3.21}$$

following Step 4 of Method-I,
Else consider $B_i = B_{i,1}$ as a T1 MF.
*End—For*;

**Step 10**: Determine the pair of IT2FS for MFVTS $\widetilde{A}_i\ or\ A_i$ and SFVTS $\widetilde{B}_j\ or\ B_j$ at current day $t-1$, whichever is available, lying in partitions $P_i$ and $Q_i$ respectively and also determine the IT2FS for MFTVS $\widetilde{A}_k\ or\ A_k$ at next day $t$, and save the fuzzy relational mapping:$A_i, B_j \to A_k\ or\ A_i, \widetilde{B}_j \to A_k\ or\ \widetilde{A}_i, B_j \to A_k\ or\ A_i, \widetilde{B}_j \to \widetilde{A}_k\ or$ $\widetilde{A}_i, B_j \to \widetilde{A}_k\ or\ \widetilde{A}_i, \widetilde{B}_j \to A_k\ or\ A_i, B_j \to \widetilde{A}_k\ or\ \widetilde{A}_i, \widetilde{B}_j \to \widetilde{A}_k$, whichever is applicable, in a list $L$.

**Step 11**: Given the variation value for MFVTS $v'_M$ and SFVTS $v'_S$ on day $(t-1)$, we first check whether the measurements $v'_M$ and $v'_S$, belonging to partitions $P_m P_m$ and $Q_s$ respectively, have membership functions lying in the antecedent of the pre-constructed rules. In case MFs corresponding to any one the linguistic variables $v'_M$ and $v'_S$ is absent from the pre-constructed rules, we take the predicted variation (of close price) as the centre of the partition $P_m$. However, if MFs for both the linguistic variables $v'_M$ and $v'_S$ are present in the antecedent of any rule, we use three distinct types of fuzzy reasoning as applicable. The general structure of fuzzy rules and reasoning mechanism are given in Table 3.1.

Let $v_K$ be the linguistic variable denoting variation in MFVTS at day t. In case the MFs of $v'_M, v'_S$ and $v_K$ are found as type-1 memberships in one of the pre-defined rules, then the reasoning is purely type-1 reasoning. In case the MFs of the above linguistic variables appear as IT2FS, we develop a pure IT2 fuzzy reasoning. For all other cases, we use mixed reasoning, where both type-1 and IT2 MFs are used jointly in the same rule and reasoning procedure. Figure 3.2a, b geometrically explains the inference generating mechanism introduced in Table 3.1.

**Step 12**: After the resulting inference is obtained, we defuzzify the inference using IT2 or T1 defuzzification, which one is applicable in Table 3.1. The defuzzified value represents the predicted variation at day t.

**Table 3.1** Fuzzy Inference rules generations for various cases evolved

| Cases | Rules | Fuzzy inference generation | Defuzzification |
|---|---|---|---|
| 1. All antecedents and consequent are T1FS | $A_i, B_j \rightarrow A_k$ | $\mu_{A'_k}(v_M(t)) = \text{Min}\,[d, \mu_{A_k}(v_M(t))]$,<br>where,<br>$d = \text{Min}[\mu_{A_i}(v'_M), \mu_{B_j}(v'_S)]$, | $c^x = \frac{\sum_{c(t_i)=-\infty}^{\infty} \mu_{A_x}(c(t_i)).c(t_i)}{\sum_{c(t_i)=-\infty}^{\infty} \mu_{A_x}(c(t_i))}$ |
| 2(a). One antecedent is T1FS(MFVS), another one is IT2FS(SFVS) and consequent is T1FS(MFVS) | $A_i, \widetilde{B}_j \rightarrow A_k$ | $\mu_{A'_k}(v_M(t)) = \text{Min}[f, \mu_{A_k}(v_M(t))]$,<br>where,<br>$umf = \text{Min}[\mu_{A_i}(v'_M), UMF_j(v'_S)]$,<br>$lmf = \text{Min}[\mu_{A_i}(v'_M), LMF_j(v'_S)]$,<br>$f = \text{Max}[umf, lmf]$ | Same as Case-1. |
| 2(b). One antecedent is T1FS(SFVS), another one is IT2FS(MFVS) and consequent is T1FS(MFVS) | $\widetilde{A}_i, B_j \rightarrow A_k$ | $umf = \text{Min}[UMF_i(v'_M), \mu_{B_j}(v'_S)]$<br>$lmf = \text{Min}[LMF_i(v'_M), \mu_{B_j}(v'_S)]$<br>$f = \text{Max}[umf, lmf]$<br>$\mu_{A'_k}(v_M(t)) = \text{Min}[f, \mu_{A_k}(v_M(t))]$, | |
| 3(a). One antecedent is T1FS(MFVS), another one is IT2FS(SFVS) and consequent is IT2FS(MFVS) | $A_i, \widetilde{B}_j \rightarrow \widetilde{A}_k$ | $umf = \text{Min}[\mu_{A_i}(v'_M), UMF_j(v'_S)]$<br>$lmf = \text{Min}[\mu_{A_i}(v'_M), LMF_j(v'_S)]$<br>$UMF'_k = \text{Min}[umf, UMF_k]$<br>$LMF'_k = \text{Min}[lmf, LMF_k]$ | Following (3.2), (3.3) and (3.4) |
| 3(b). One antecedent is T1FS(SFVS), another one is IT2FS(MFVS) and consequent is IT2FS(MFVS) | $\widetilde{A}_i, B_j \rightarrow \widetilde{A}_k$ | $umf = \text{Min}[UMF_i(v'_M), \mu_{B_j}(v'_S)]$<br>$lmf = \text{Min}[LMF_i(v'_M), \mu_{B_j}(v'_S)]$<br>$UMF'_k = \text{Min}[umf, UMF_k]$<br>$LMF'_k = \text{Min}[lmf, LMF_k]$ | |
| 4. Both the antecedents are IT2FS and consequent is T1FS | $\widetilde{A}_i, \widetilde{B}_j \rightarrow A_k$ | $umf = \text{Min}[UMF_i(v'_M), UMF_j(v'_S)]$<br>$lmf = \text{Min}[LMF_i(v'_M), LMF_j(v'_S)]$<br>$f = \text{Max}[umf, lmf]$<br>$\mu_{A'_k}(v_M(t)) = \text{Min}[f, \mu_{A_k}(v_M(t))]$ | Same as Case-1. |
| 5. Both the antecedents are T1FS and consequent is IT2FS | $A_i, B_j \rightarrow \widetilde{A}_k$ | $e = \text{Min}[\mu_{A_i}(v'_M),\ \mu_{B_i}(v'_S)]$<br>$UMF'_k = Min[e, UMF_k]$<br>$LMF'_k = \text{Min}[e, LMF_k]$ | Following (3.2), (3.3) and (3.4) |
| 6. Both the antecedents and consequent are IT2FS | $\widetilde{A}_i, \widetilde{B}_j \rightarrow \widetilde{A}_k$ | Following Method II | Following Method II |

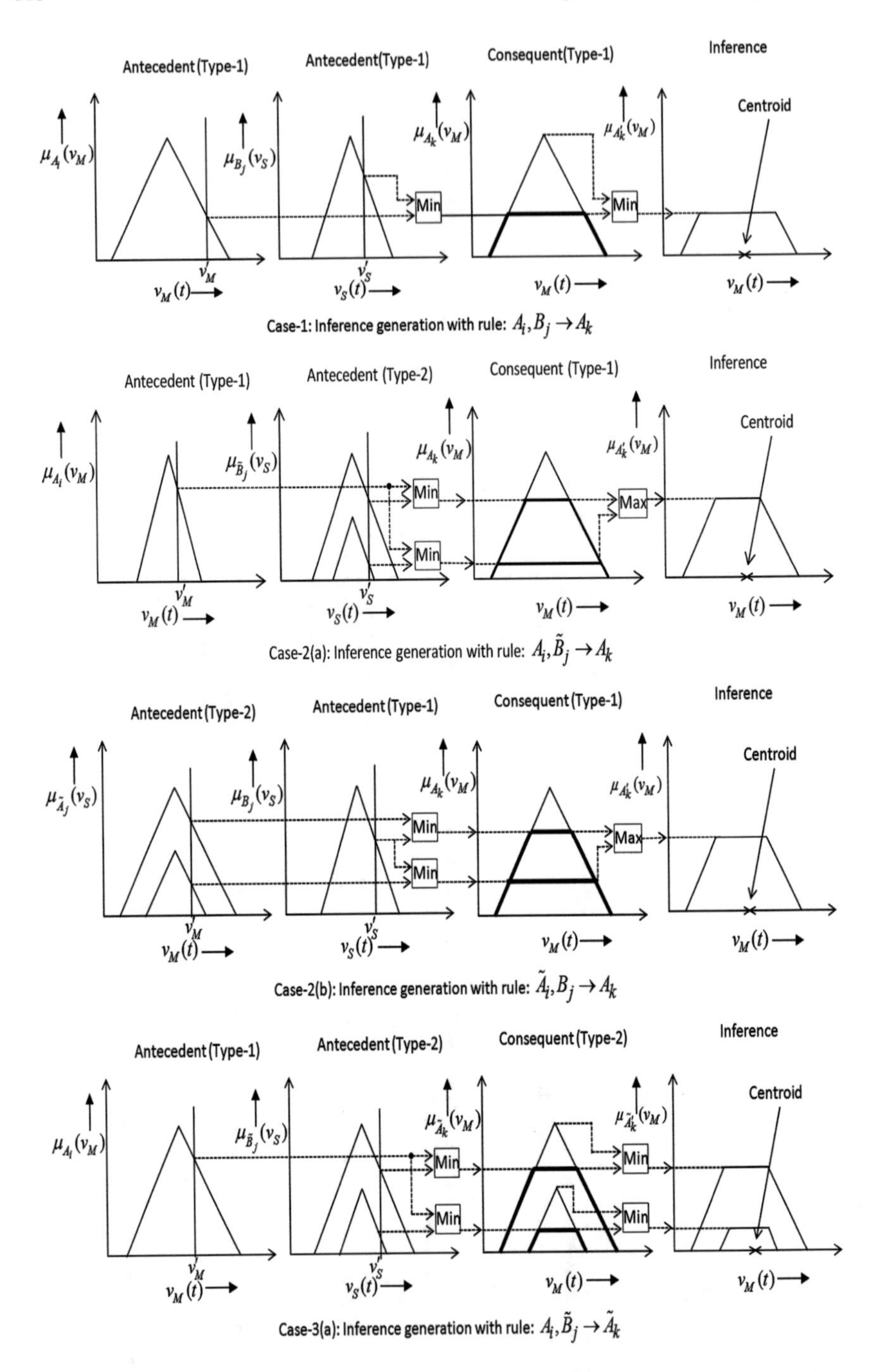


**Fig. 3.2** **a** Inference generation with T1/IT2 antecedent and consequent pair, **b** inference generation with T1/T2 antecedent consequent pair

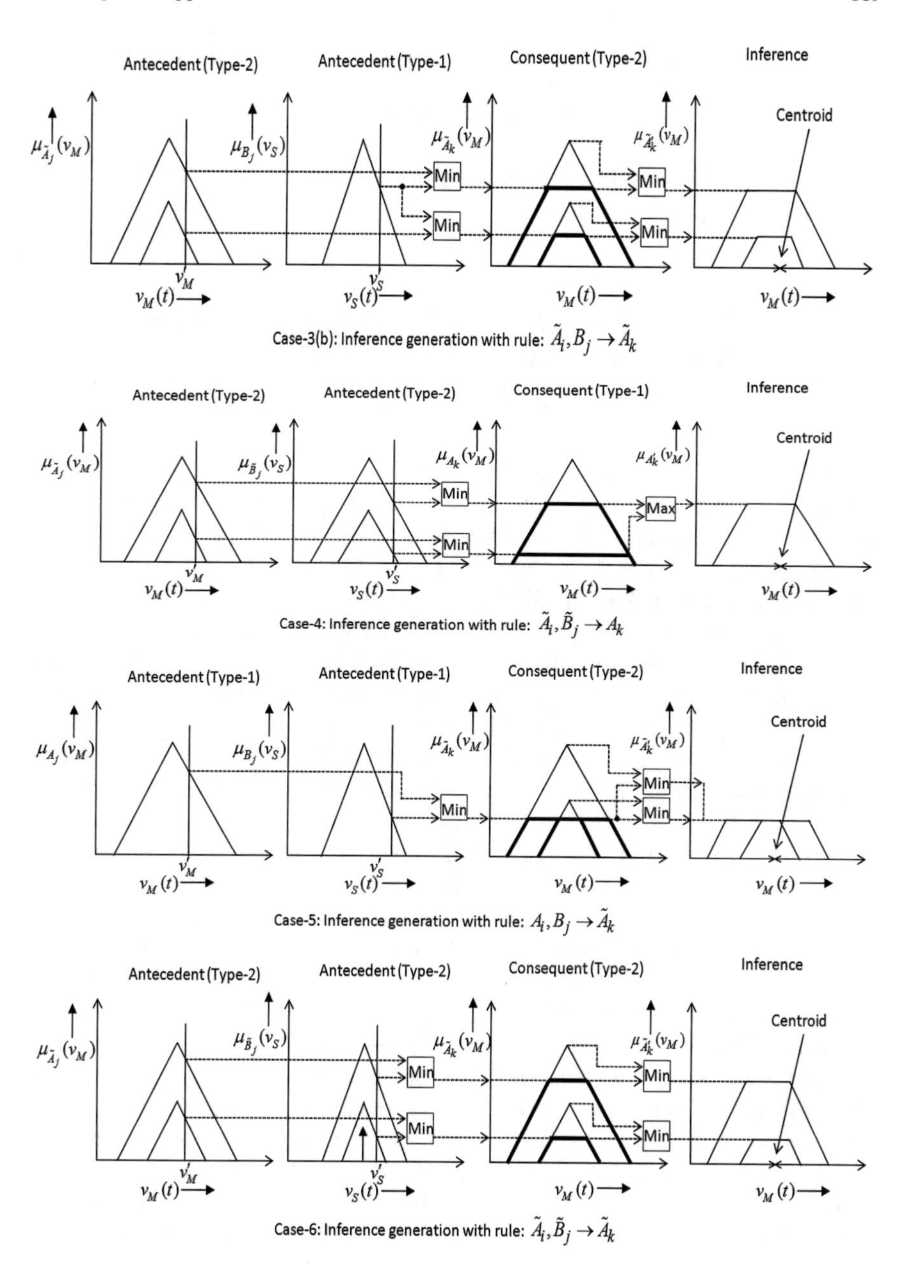


**Fig. 3.2** (continued)

**Step 13**: The Predicted price at day *t* i.e., c(t) is evaluated by the following step:

$$c(t) = c(t-1) + v(t),$$

where $v(t) = \bar{c}$ the defuzzified signal amplitude, representing the predicted variation for the next day t.

### *3.3.4 Method-IV: Adaptation of Membership Function in Method III to Handle Dynamic Behaviour of Time-Series [47–52]*

In Method-III, we constructed membership functions from the time-series data of first 10 calendar months and attempted to predict the time-series daily for the next two months. Our experience with time-series prediction [6] reveals that occasionally such prediction suffers, perhaps, due to structural changes in the series during the last two calendar months. Method-IV overcomes this problem by arranging adaptation of MFs after a regular interval of 15 days starting from 1st November. Such adaptation in MF improves the prediction accuracy, as indicated by RMSE [4], for the next 15 days after the MFs are adapted. This section proposes a novel approach to optimally adapt the MFs used in the prediction rules adopted in Method III. In order to accomplish the tuning of MFs, we attempt to optimally select the base-width of the isosceles triangular MFs. Any traditional meta-heuristic algorithm could serve the optimization problem. We, however, select the well-known Differential Evolution (DE) algorithm for its low computational overhead, simplicity and above all our familiarity with the algorithm [51, 52].

While employing DE, we consider base widths of the trial solutions in DE as unknown with a motivation to minimize J, the fitness function, as introduced below. In the proposed algorithm, trial solutions representing base widths of the MFs are randomly initialized within $\pm 30\%$ of their actual values used for non-adaptive prediction. The mutation, crossover and selection steps are like those of standard DE. The fitness function evaluation in the selection step, however, requires running the prediction algorithm daily for fifteen days for different settings of trial solutions. The algorithm is iterated until the convergence in fitness measure is attained. Several methods for testing convergence of the algorithm are available in the literature. We here considered 'no further improvement in average fitness of the population' as the metric to test the convergence criterion. The steps of the adaptation algorithm are outlined below.

**Step 1**: Initialize population of trial solution $TS_i$, where each trial solution includes the base width of type-1 MFs present in the rules used for prediction from *i*th to $(i+1)$th day randomly from a range of $\pm 30\%$ of their original base widths.

**Step 2**: For each day $i$, starting from the beginning of the 11-th month and continuing up to the middle of the 11-th month, do evaluate $RMSE_i = |(AV_i - FV_i)|$, where $AV_i$ and $FV_i$ are actual close price value and forecasted close price value on $i$th day respectively and hence determine the metric $J = \sum_{i=1}^{i=15} RMSE_i$, where $i$ = 1 to 15 corresponding to training period of first day of November and the 15th day of November, considering 15 calendar days in the month of November.

**Step 3**: Considering J as the fitness value of the trial solutions, change the trial solutions by undertaking Mutation, Crossover and select the trial solutions for the next generation (by undertaking the Selection step of DE) [47–52].

**Step 4**: Repeat from step 2 until the difference between average fitness of two consecutive generation is lower than a predefined very small positive real number $\varepsilon$.

**Step 5**: Report the best trial solution containing the optimal settings of base widths of the T1 MFs used in the rules.

**Step 6**: Loop through step 2 with redefined value of i increased by 15 days until i attains 31st December of the calendar year. This is adopted by the principles outlined below.

The type-1 MFs are adapted using the above algorithm for every fifteen days to predict for the next fifteen days. In our study, we determined prediction rules and the membership functions from the time-series data for the first ten months (1st January to 31st October). Then we adapted the type-1 MFs using the prediction results obtained by method III for the next fifteen days (November 1–15). Next we go for daily prediction of the time-series for the period: November 16 to November 30 using the adapted MFs obtained previously.

In order to predict the time-series for the period December 1 to 15, we first need to adapt type-1 MFs for the period (November 16–31) with the predicted result obtained during this period, and then go for the step of prediction. Lastly, to predict the time series for the last fifteen days in the year, we adapt MFs for the first fifteen days in December and perform prediction with the adapted type-1 MFs.

## 3.4 Experiments

### 3.4.1 Experimental Platform

We performed the experiment using MATLAB 2012b under operating system as WINDOWS-7 running on a IBM personal computer with Core i5 processor with system clock of 3.60 G-Hz frequency and system RAM of 8 GB.

### 3.4.2 Experimental Modality and Results

**Policies considered**: Experiments are performed on TAIEX stock data for the duration of 1999–2004 [46], considering January 1 to October 31 as the training period and November 1 to December 31 as the testing period, irrespective of any methods. Here, we transform the actual time-series into a variation series by taking the differences of two consecutive close price data.

The close price data for both main factor and secondary factors are obtained for the period 1990–2004 from the website [46]. The training session includes all trading days. In case all the trading days of secondary factors do not coincide with those of the main factor, we adopt policies for the following two cases.

Case-I: Let us consider two sets A and B, representing respectively the dates of trading in main and secondary factor time-series. If the set difference: A − B is non-null, we retain the close price of previous trading days in secondary factor over the missing (subsequent) days.

Case-II: If the set difference: B − A is non-null, the following policy is adopted as applicable. In case the main factor has missing trading days due to holidays and/or other local factors, we simply omit those days for training. Further, in the trading of next day of main factor time-series, the influence of the last day of trading in secondary factor closing price would be considered. After completion of the training, prediction rules (also called Fuzzy Logic Implications (FLI)) are saved for prediction of the main factor time series.

**MF Selection**: From our previous experience [6], we note that triangular type-1 MFs yield better prediction accuracy than those obtained by Gaussian MFs. This prompted us to take up the present study with triangular MFs. In method I to III, we fix the type-1 triangular MFs once only. However, in method IV, we need to adapt the parameters of the triangular MFs after the training session is over. Such adaptation is required to improve RMSE in the prediction phase performed during the last two months of the calendar year.

In this section, we attempt to compare the relative performance of the proposed four methods with existing techniques for stock index time-series prediction. We used root mean square error (RMSE) as the metric for this study. All the experiments are undertaken with the TAIEX stock data as the main factor time series. Method-I utilizes only the MFTS. Method-II to IV however used NASDAQ as the secondary factor time-series. We used 28 well-known techniques to compare the relative performance of the proposed techniques with them.

Table 3.2 provides the list of rules obtained from the 2003 TAIEX main-factor time series using Method-II, III and IV with NASDAQ as secondary factor time-series. The rule selection is performed for the said time-series for the first 10 months. The rules obtained by Method-I are simply the implication rules without $B_j, \forall j$. In addition to the above, we need to adapt MFs in Method-IV at 15 days' interval and validate it for the next 15 days. We repeat it starting from

**Table 3.2** Fuzzy logical implications for 2003

| |
|---|
| $A_{12}, B_1 \rightarrow A_{13}$ |
| $A_9, B_2 \rightarrow A_{12}$; $A_{11}, B_2 \rightarrow A_9$; $A_{12}, B_2 \rightarrow A_{13}$; $A_{13}, B_2 \rightarrow A_{13}$; $A_{14}, B_2 \rightarrow A_{12}, A_{12}$ |
| $A_3, B_3 \rightarrow A_9$; $A_7, B_3 \rightarrow A_{11}$; $A_9, B_3 \rightarrow A_7$; $A_{11}, B_3 \rightarrow A_{13}$; $A_{12}, B_3 \rightarrow A_{11}, A_{11}, A_{12}, A_{13}, A_{14}, A_{14}$;<br>$A_{13}, B_3 \rightarrow A_{11}, A_{11}, A_{11}, A_{12}, A_{14}$; $A_{14}, B_3 \rightarrow A_{12}$; $A_{15}, B_3 \rightarrow A_{12}, A_{13}$ |
| $A_7, B_4 \rightarrow A_{14}$; $A_8, B_4 \rightarrow A_{14}, A_{14}$; $A_9, B_4 \rightarrow A_{14}$; $A_{10}, B_4 \rightarrow A_{14}$; $A_{11}, B_4 \rightarrow A_{13}$<br>$A_{12}, B_4 \rightarrow A_{12}, A_{12}, A_{12}, A_{13}$; $A_{13}, B_4 \rightarrow A_3, A_{11}, A_{12}, A_{13}, A_{13}$; $A_{15}, B_4 \rightarrow A_{13}$ |
| $A_{10}, B_5 \rightarrow A_{12}$; $A_{11}, B_5 \rightarrow A_{11}$; $A_{12}, B_5 \rightarrow A_{11}, A_{12}, A_{13}, A_{13}, A_{13}, A_{13}, A_{14}, A_{14}, A_{15}$; $A_{13}, B_5 \rightarrow A_{12}$;<br>$A_{14}, B_5 \rightarrow A_{11}, A_{12}, A_{12}, A_{15}$; $A_{15}, B_5 \rightarrow A_8, A_{11}$; $A_{16}, B_5 \rightarrow A_{13}$; $A_{17}, B_5 \rightarrow A_{13}$ |
| $A_9, B_6 \rightarrow A_{13}$; $A_{10}, B_6 \rightarrow A_8, A_{12}$; $A_{11}, B_6 \rightarrow A_{11}, A_{11}, A_{11}, A_{11}, A_{12}, A_{12}, A_{13}, A_{14}, A_{14}, A_{15}$; $A_{15}, B_6 \rightarrow A_{13}$;<br>$A_{12}, B_6 \rightarrow A_{10}, A_{11}, A_{12}, A_{12}, A_{13}, A_{13}, A_{15}, A_{15}, A_{19}$; $A_{14}, B_6 \rightarrow A_{10}, A_{11}, A_{12}, A_{12}, A_{15}$; $A_{16}, B_6 \rightarrow A_{12}$ |
| $A_6, B_7 \rightarrow A_{14}$; $A_9, B_7 \rightarrow A_{13}$; $A_{10}, B_7 \rightarrow A_9$; $A_{11}, B_7 \rightarrow A_{10}, A_{12}, A_{12}, A_{13}, A_{13}, A_{14}, A_{15}, A_{16}$<br>$A_{12}, B_7 \rightarrow A_9, A_{10}, A_{10}, A_{12}, A_{12}, A_{12}, A_{12}, A_{12}, A_{12}, A_{14}$; $A_{14}, B_7 \rightarrow A_{11}, A_{12}$; $A_{19}, B_7 \rightarrow A_{12}$;<br>$A_{13}, B_7 \rightarrow A_{10}, A_{11}, A_{11}, A_{11}, A_{11}, A_{12}, A_{12}, A_{12}, A_{13}, A_{13}, A_{13}, A_{13}, A_{13}, A_{13}, A_{14}, A_{14}, A_{14}, A_{14}, A_{15}$ |
| $A_{10}, B_8 \rightarrow A_{12}$; $A_{11}, B_8 \rightarrow A_{16}$; $A_{12}, B_8 \rightarrow A_{12}, A_{12}, A_{13}, A_{13}, A_{13}, A_{13}, A_{14}$;<br>$A_{13}, B_8 \rightarrow A_{11}, A_{12}, A_{12}, A_{12}, A_{13}, A_{14}, A_{16}$; $A_{14}, B_8 \rightarrow A_{12}, A_{12}, A_{12}, A_{12}, A_{12}, A_{12}, A_{13}$<br>$A_{15}, B_8 \rightarrow A_{12}, A_{13}$; $A_{16}, B_8 \rightarrow A_5, A_{13}$ |
| $A_{12}, B_9 \rightarrow A_{11}, A_{12}, A_{12}, A_{13}, A_{13}, A_{13}, A_{13}, A_{14}$;<br>$A_{13}, B_9 \rightarrow A_{12}, A_{13}, A_{13}, A_{14}, A_{14}, A_{15}, A_{16}$; $A_{14}, B_9 \rightarrow A_{11}, A_{19}$;<br>$A_{15}, B_9 \rightarrow A_{11}, A_{12}, A_{14}$; $A_{16}, B_9 \rightarrow A_{13}$ |
| $A_9, B_{10} \rightarrow A_{14}$; $A_{11}, B_{10} \rightarrow A_{13}$; $A_{12}, B_{10} \rightarrow A_{12}, A_{17}$; $A_{13}, B_{10} \rightarrow A_{13}$;<br>$A_{14}, B_{10} \rightarrow A_{12}, A_{14}, A_{16}$; $A_{19}, B_{10} \rightarrow A_{13}$ |
| $A_{11}, B_{11} \rightarrow A_9$; $A_{12}, B_{11} \rightarrow A_{13}$; $A_{13}, B_{11} \rightarrow A_{14}, A_{15}$;<br>$A_{14}, B_{11} \rightarrow A_{13}, A_{15}, A_{15}$; $A_{17}, B_{11} \rightarrow A_{14}$ |
| $A_{11}, B_{12} \rightarrow A_{12}$ |

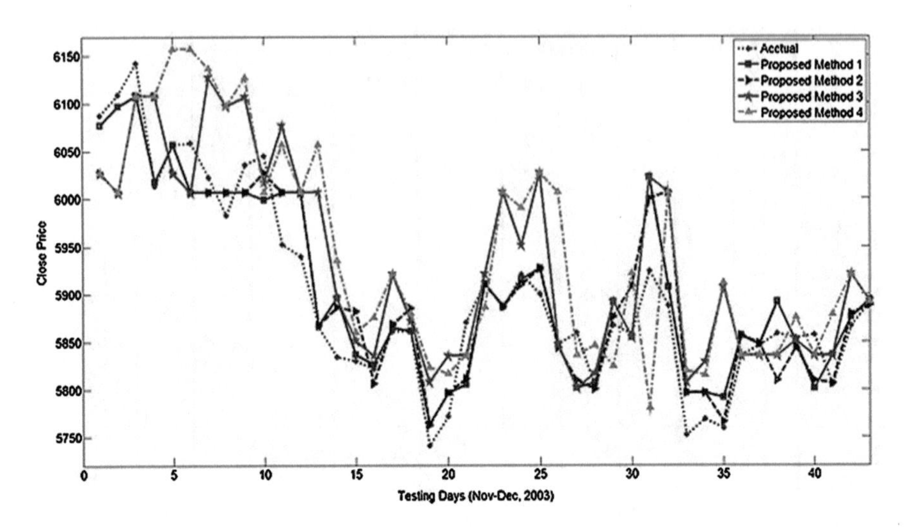


**Fig. 3.3** Forecasted TAIEX of the months November and December 2003 using Triangular MFs, (1) Actual TAIEX; (2) Forecasted data using proposed method I; (3) Forecasted data using proposed method II; (4) Forecasted data using proposed method III; (5) Forecasted data using proposed method IV

**Table 3.3** Comparison of RMSEs obtained by the proposed techniques

| Methods | Years | | | | | | |
|---|---|---|---|---|---|---|---|
| | 1999 | 2000 | 2001 | 2002 | 2003 | 2004 | Mean |
| 1. Proposed method I(1) | 175.74 | 192.75 | 230.45 | 181.26 | 144.2603 | 197.34 | 241.3001 |
| 2. Proposed method II(2) | 122.493 | 169.314 | 169.614 | 129.547 | 119.8202 | 132.719 | 152.2512 |
| 3. Proposed method III(3) | 83.141 | 116.471 | 101.423 | 56.756 | 43.71476 | 49.471 | 75.1628 |
| 4. **Proposed method IV(4)** | **74.541** | **110.145** | **94.127** | **50.441** | **33.75384** | **39.065** | **67.0121** |

October to December with 15 days' interval. Using the prediction rules stated in Table 3.2, we plot the predicted time-series value by the proposed four methods along with the actual time-series in Fig. 3.3 and evaluate the RMSE for all the four existing works. The results are listed in Table 3.4. It is apparent from Table 3.4 that Method-IV yields the best performance with respect to RMSE, followed by Method-III, II and I in order. It is also noted that our proposed four methods outperform existing techniques with RMSE as the key metric (Table 3.3).

**Table 3.4** Comparison of RMSE obtained by the proposed technique with existing techniques

| Methods | Years | | | | | | |
|---|---|---|---|---|---|---|---|
| | 1999 | 2000 | 2001 | 2002 | 2003 | 2004 | Mean |
| 1. Huarng et al. [30] (using NASDAQ) | NA | 158.7 | 136.49 | 95.15 | 65.51 | 73.57 | 105.88 |
| 2. Huarng et al. [30] (using Dow Jones) | NA | 165.8 | 138.25 | 93.73 | 72.95 | 73.49 | 108.84 |
| 3. Huarng et al. [30] (using $M_{1b}$) | NA | 160.19 | 133.26 | 97.1 | 75.23 | 82.01 | 111.36 |
| 4. Huarng et al. [30] (using NASDAQ and $M_{1b}$) | NA | 157.64 | 131.98 | 93.48 | 65.51 | 73.49 | 104.42 |
| 5. Huarng et al. [30] (using Dow Jones and $M_{1b}$) | NA | 155.51 | 128.44 | 97.15 | 70.76 | 73.48 | 105.07 |
| 6. Huarng et al. [30] (using NASDAQ,Dow Jones and $M_{1b}$) | NA | 154.42 | 124.02 | 95.73 | 70.76 | 72.35 | 103.46 |
| 7. Chen et al. [21, 31, 32] | 120 | 176 | 148 | 101 | 74 | 84 | 117.4 |
| 8. U_R model [31, 32] | 164 | 420 | 1070 | 116 | 329 | 146 | 374.2 |
| 9. U_NN model [31, 32] | 107 | 309 | 259 | 78 | 57 | 60 | 145.0 |
| 10. U_NN_FTS model [16, 31, 32] | 109 | 255 | 130 | 84 | 56 | 116 | 125.0 |
| 11. U_NN_FTS_S model [19, 31, 32] | 109 | 152 | 130 | 84 | 56 | 116 | 107.8 |
| 12. B_R model [31, 32] | 103 | 154 | 120 | 77 | 54 | 85 | 98.8 |
| 13. B_NN model [31, 32] | 112 | 274 | 131 | 69 | 52 | 61 | 116.4 |
| 14. B_NN_FTS model [31, 32] | 108 | 259 | 133 | 85 | 58 | 67 | 118.3 |
| 15. B_NN_FTS_S model [31, 32] | 112 | 131 | 130 | 80 | 58 | 67 | 96.4 |
| 16. Chen et al. [4] (using Dow Jones) | 115.47 | 127.51 | 121.98 | 74.65 | 66.02 | 58.89 | 94.09 |
| 17. Chen et al. [4] (using NASDAQ) | 119.32 | 129.87 | 123.12 | 71.01 | 65.14 | 61.94 | 95.07 |
| 18. Chen et al. [4] (using $M_{1b}$) | 120.01 | 129.87 | 117.61 | 85.85 | 63.1 | 67.29 | 97.29 |
| 19. Chen et al. [4] (using NASDAQ and Dow Jones) | 116.64 | 123.62 | 123.85 | 71.98 | 58.06 | 57.73 | 91.98 |
| 20. Chen et al. [4] (using Dow Jones and $M_{1b}$) | 116.59 | 127.71 | 115.33 | 77.96 | 60.32 | 65.86 | 93.96 |
| 21. Chen et al. [4] (using NASDAQ and $M_{1b}$) | 114.87 | 128.37 | 123.15 | 74.05 | 67.83 | 65.09 | 95.56 |
| 22. Chen et al. [4] (using NASDAQ,Dow Jones and $M_{1b}$) | 112.47 | 131.04 | 117.86 | 77.38 | 60.65 | 65.09 | 94.08 |
| 23. Karnik-Mendel [41] induced stock prediction | 116.60 | 128.46 | 120.62 | 78.60 | 66.80 | 68.48 | 96.59 |

(continued)

**Table 3.4** (continued)

| Methods | Years | | | | | | |
|---|---|---|---|---|---|---|---|
| | 1999 | 2000 | 2001 | 2002 | 2003 | 2004 | Mean |
| 24. Chen et al. [5] (using NASDAQ, Dow Jones and $M_{1b}$) | 101.47 | 122.88 | 114.47 | 67.17 | 52.49 | 52.84 | 85.22 |
| 25. Chen et al. [53] | 87.67 | 125.34 | 114.57 | 76.86 | 54.29 | 58.17 | 86.14 |
| 26. Cai et al. [54] | 102.22 | 131.53 | 112.59 | 60.33 | 51.54 | 50.33 | 84.75 |
| 27. Chen [55] | NA | 108 | 88 | 60 | 42 | NA | 74.5 |
| 28. Askaria et al. [56] | 57.94 | 83.791 | 43.3371 | 43.757 | 30.8842 | 38.138 | 49.65 |
| **29. Proposed method 4 (IV)** | **74.541** | **110.14** | **94.127** | **50.441** | **33.753** | **39.065** | **67.012** |

## 3.5 Conclusion

In this chapter, we proposed four incrementally improved methods of economic time-series prediction using interval type-2 fuzzy sets, and compared their relative performance with a set of 28 well-known existing techniques with RMSE as the key metric. The first method attempted to predict economic time-series using classical IT2FS. The second method attempted to use both main- and secondary factor time-series for the prediction of the main factor time-series using IT2FS. The third method deals with insufficient number of data points in training phase to construct MFs. The fourth method adapts the MFS to handle the dynamic behavior of the time-series.

We here undertook experiments with triangular MFs. Gaussian MFs are not intentionally used as we have noted relatively worse performance of Gaussian MFs in our previous work in comparison to triangular MFs. Experiments undertaken confirmed that there is a chronological improvement in performance based on the use of improved strategies in the methods proposed. The RMSE metric computed over 6 years: 1999–2004, for example, indicates a steady decrease for the four methods. For example, for methods 1, 2, 3 and 4, the measured RMSE in the year 2003 are obtained as 145, 120, 44 and 34 respectively.

An analysis of results using RMSE as the metric further indicates that the proposed methods outperform the existing techniques on stock index prediction by a considerable margin ($\geq 15\%$). Out of the four proposed methods, the method 4 with provision for its adaptation of MFs yields the best performance following the prediction of TAIEX stock data for the period of 1999–2004 with NASDAQ as the secondary index.

### Exercise

1. Consider the fuzzy prediction rule: If the secondary factor is $B_3$ on the t-th day and main factor is $A_2$ then the close price at day $(t - 1)$ is $A_3$. Given triangular MFs of $B_3$, $A_2$ and $A_3$ is indicated below. If secondary factor is 2K and main factor is 2.5K, then compute the IT2 fuzzy inference (Fig. 3.4).
2. One simple form of defuzzification is to represent the IT2FS by a type 1 fuzzy set. This is one form type reduction. After transformation of the IT2FS given in Fig. 3.5 into type -1 fuzzy set, evaluate the centroid of the type-1 fuzzy set.
   **Hints:** Compare the type-reduced fuzzy set as follows:
   In Fig. 3.6 OAB is the derived type-1 MF centroid of OAB.

## Appendix 3.1: Differential Evolution Algorithm [36, 48–50]

The classical Differential Evolution (DE) algorithm consists of the four fundamental steps—initialization of population vectors, mutation, crossover or recombination, and selection. The steps are given below.

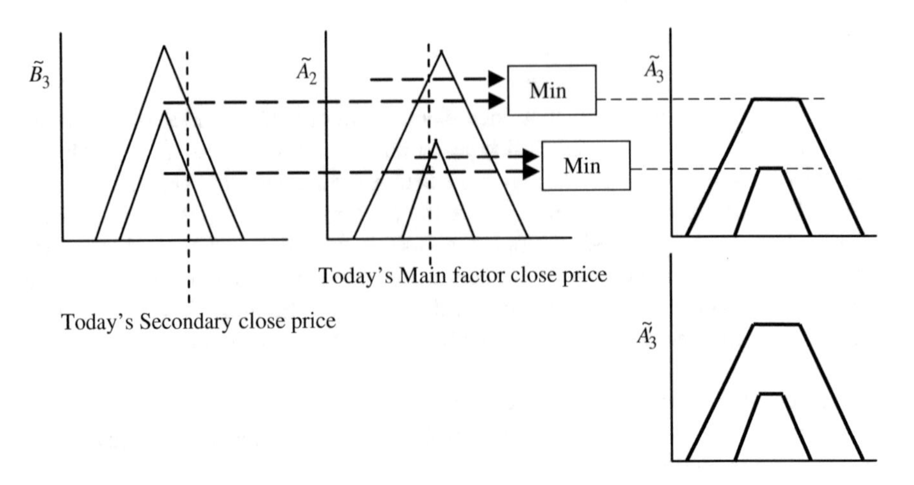


**Fig. 3.4** Figure for problem 1

**Fig. 3.5** Figure for problem 2

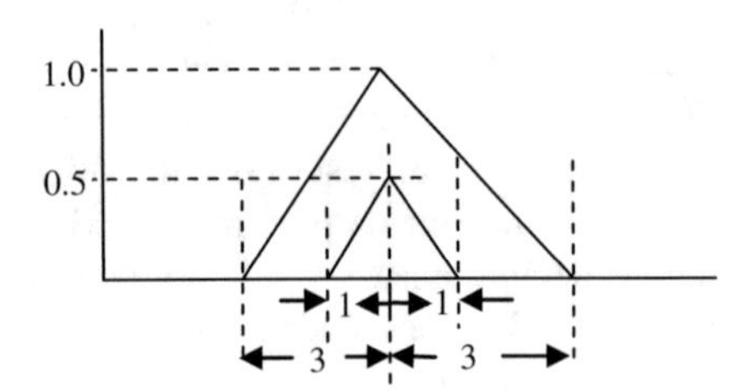


**Fig. 3.6** Figure for problem 2

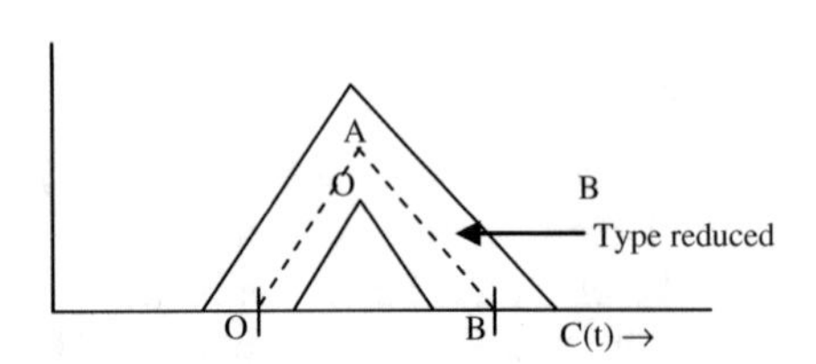


I. Initialize the generation number $t = 0$ and randomly initialize a population of $NP$ individuals $\_Pt = \{x_1(t), x_2(t), \ldots, x_{NP}(t)\}$ with $x_i(t) = \{x_{i,1}(t), x_{i,2}(t),$ 2026, $x_{i,D}(t)\}$ and each individual uniformly distributed in the range $[x_{min}, x_{max}]$, where $\mathrm{x}_{\min} = \{\mathrm{x}_{\min,1}, \mathrm{x}_{\min,2}, \ldots, \mathrm{x}_{\min,D}\}$ and $x_{max} = \{\mathrm{x}_{\max,1}, \mathrm{x}_{\max,2}, \ldots, \mathrm{x}_{\max,D}\}$ for $i = [1, 2, \ldots, NP]$.

II. **While** stopping criterion is not reached, **do Begin**
**For** $i = 1$ **to** $NP$

(a) **Mutation**: Obtain a donor vector $V(t) = \{\mathrm{v}_{i,1}(\mathrm{t}), \mathrm{v}_{i,2}(\mathrm{t}), \ldots, \mathrm{v}_{i,D}(\mathrm{t})\}$ corresponding to the $i$-th target vector $x_i(t)$ by the following scheme $V_i(t) = x_{r1}(t) + F(x_{r2}(t) - x_{r3}(t))$ where $r1$, $r2$ *and* $r3$ are randomly chosen distinct integers in [1, NP].

(b) **Crossover**: Generate trial vector $U_i(t) = \{u_{i,1}(t), u_{i,2}(t), \ldots, u_{i,D}(t)\}$ for the $i$-th target vector $x_i(t)$ by binomial crossover, given by

$$u_{i,j}(t) = v_{i,j}(t) \ if \ rand\ (0,1) < Cr, \text{ a fixed crossover rate}$$
$$= x_{i,j}(t) \ otherwise.$$

(c) **Selection**: Evaluate the trial vector $u_i(t)$

**If** $f(U_i(t)) < f(x_i(t))$,
**Then** $x_i(t+1) = U_i(t)$
**End if**;

**End for;**
(d) Increase the counter value t = t + 1.
**End while;**

In the above algorithm, we introduced two parameters, called crossover rate $C_r$ and scale factor F. They are defined once for all at the beginning of the program. Typically $C_r$ lies in [0, 2] and has been fixed as 1.2 in the program. Similarly, typical range of F to attain convergence is [0, 2], and we fixed it as 0.7. In addition, selection of terminating condition is an important issue. It could be defined in many alternative ways, such as (1) fixing upper bound of program iterations, (2) fixing an error bound to the best-fit member in the population and best fit average fitness over each iterations or any one of the above whichever occurs earlier.

## References

1. Engineering Statistics and book [Online], http://www.itl.nist.gov/div898/handbook/pmc/section4/pmc41.htm
2. Chen, S. M., & Chen, C. D. (2011). Handling forecasting problems based on high-order fuzzy logical relationships. *Expert Systems with Applications, 38*(4), 3857–3864.
3. Chen, S. M., & Tanuwijaya, K. (2011). Fuzzy forecasting based on high-order fuzzy logical relationships and automatic clustering techniques. *Expert Systems with Applications, 38*(12), 15425–15437.
4. Chen, S. M., & Chen, C. D. (2011). TAIEX forecasting based on fuzzy time series and fuzzy variation groups. *IEEE Transactions on Fuzzy Systems, 19*(1), 1–12.
5. Chen, S. M, Chu, H. P., & Sheu, T. W. (2012). TAIEX Forecasting using fuzzy time series and automatically generated weights of multiple factors. *IEEE Transactions on Systems, Man, and Cybernetics-Part A: Systems and Humans, 42*(6), 1485–1495.
6. Bhattacharya, D., Konar, A., & Das, Pratyusha. (2016). Secondary factor induced stock index time-series prediction using self-adaptive interval type-2 fuzzy sets. *Neurocomputing, 171,* 551–568.
7. Wu, C. L., & Chau, K. W. (2013). Prediction of rainfall time series using modular soft computing methods. *Engineering Applications of Artificial Intelligence, 26*, 997–1007.
8. Barnea, O., Solow, A. R., & Stonea, L. (2006). On fitting a model to a population time series with missing values. *Israel Journal of Ecology & Evolution, 52,* 1–10.

9. Chen, S. M., & Hwang, J. R. (2000). Temperature prediction using fuzzy time series. *IEEE Transactions on Systems, Man, and Cybernetics. Part B, Cybernetics, 30*(2), 263–275.
10. Song, Q., & Chissom, B. S. (1993). Fuzzy time series and its model. *Fuzzy Sets and Systems, 54*(3), 269–277.
11. Song, Q., & Chissom, B. S. (1993). Forecasting enrollments with fuzzy time series—Part I. *Fuzzy Sets and Systems, 54*(1), 1–9.
12. Song, Q., & Chissom, B. S. (1994). Forecasting enrollments with fuzzy time series—Part II. *Fuzzy Sets and Systems, 62*(1), 1–8.
13. Song, Q. (2003). A note on fuzzy time series model selection with sample autocorrelation functions. *Cybernetics and Systems, 34*(2), 93–107.
14. Wong, H. L., Tu, Y. H., & Wang, C. C. (2009). An evaluation comparison between multivariate fuzzy time series with traditional time series model for forecasting Taiwan export. In *Proceedings of World Congess on Computer Science and Engineering* (pp. 462–467).
15. Wu, C. L., Chau, K. W., & Fan, C. (2010). Prediction of rainfall time series using modular artificial neural networks coupled with data-preprocessing techniques. *Journal of Hydrology, 389*(1–2), 146–167.
16. Huarng, K., & Yu, T. H. K. (2006). The application of neural networks to forecast fuzzy time series. *Physica A, 363*(2), 481–491.
17. Yu, T. H. K., & Huarng, K. H. (2010). A neural network-based fuzzy time series model to improve forecasting. *Expert Systems with Applications, 37*(4), 3366–3372.
18. Jansen, B. H., Bourne, J. R., & Ward, J. W. (1981). Autoregressive estimation of short segment spectra for computerized EEG analysis. *IEEE Transactions on Biomedical Engineering, 28*(9).
19. Lee, L. W., Wang, L. H., & Chen, S. M. (2007). Temperature prediction and TAIFEX forecasting based on fuzzy logical relationships and genetic algorithms. *Expert Systems with Applications, 33*(3), 539–550.
20. Kuo, H., Horng, S. J., Kao, T. W., Lin, T. L., Lee, C. L., & Pan, Y. (2009). An improved method for forecasting enrollments based on fuzzy time series and particle swarm optimization. *Expert Systems with Applications, 36*(3), 6108–6117.
21. Chen, S. M. (1996). Forecasting enrollments based on fuzzy time series. *Fuzzy Sets and Systems, 81*(3), 311–319.
22. Cheng, C. H., Chen, T. L., Teoh, H. J., & Chiang, C. H. (2008). Fuzzy time-series based on adaptive expectation model for TAIEX forecasting. *Expert Systems with Applications, 34*, 1126–1132.
23. Chen, T. L., Cheng, C. H., & Teoh, H. J. (2008). High-order fuzzy time-series based on multi-period adaptation model for forecasting stock markets. *Physica A, 387,* 876–888.
24. Lu, W., Yang, J., Liu, X., & Pedrycz, W. (2014). The modeling and prediction of time-series based on synergy of high order fuzzy cognitive map and fuzzy c-means clustering. *Knowledge Based Systems, 70*, 242–255.
25. Chen, S. M., & Tanuwijaya, K. (2011). Multivariate fuzzy forecasting based on fuzzy time series and automatic clustering clustering techniques. *Expert Systems with Applications, 38* (8), 10594–10605.
26. Huarng, K., Yu, H. K., & Hsu, Y. W. (2007). A multivariate heuristic model for fuzzy time-series forecasting. *IEEE Transactions on Systems, Man, and Cybernetics. Part B, Cybernetics, 37*(4), 836–846.
27. Chen, T. L., Cheng, C. H., & Teoh, H. J. (2007). Fuzzy time-series based on Fibonacci sequence for stock price forecasting. *Physica A, 380,* 377–390.
28. Jalil, A., & Idrees, M. (2013). Modeling the impact of education on the economic growth: Evidence from aggregated and disaggregated time series data of Pakistan. *Economic Modelling, 31*, 383–388.
29. Hwang, J. R., Chen, S. M., & Lee, C. H. (1998). Handling forecasting problems using fuzzy time series. *Fuzzy Sets and Systems, 100*(1–3), 217–228.

30. Wong, W. K., Bai, E., & Chu, A. W. C. (2010). Adaptive time-variant models for fuzzy-time-series forecasting. *IEEE Transactions on Systems, Man, Cybernetics-Part B: Cybernetics, 40*(6).
31. Yu, T. H. K., & Huarng, K. H. (2008). A bivariate fuzzy time series model to forecast the TAIEX. *Expert Systems with Applications, 34*(4), 2945–2952.
32. Yu, T. H. K., & Huarng, K. H. (2010). Corrigendum to "A bivariate fuzzy time series model to forecast the TAIEX. *Expert Systems with Applications, 37*(7), 5529.
33. Chen, S. M., & Chang, Y. C. (2010). Multi-variable fuzzy forecasting based on fuzzy clustering and fuzzy rule interpolation techniques. *Information Sciences, 180*(24), 4772–4783.
34. Sun, B. Q., Guo, H., Karimi, H. R., Ge, Y., & Xiong, S. (2015). Prediction of stock index futures prices based on fuzzy sets and multivariate fuzzy time series. *Neurocomputing, 151*, 1528–1536.
35. Garibaldi, J. M. (2010). Alternative forms of non-standard fuzzy sets: A discussion chapter. In *Proceedings of Type-2 Fuzzy Logic: State of the Art and Future Directions, London*, June 2010.
36. Halder, A., Konar, A., Mandal, R., Chakraborty, A., Bhowmik, P., Pal, N. R., et al. (2013). General and interval type-2 fuzzy face-space approach to emotion recognition. *IEEE Transactions On Systems, Man, And Cybernetics: Systems, 43*(3).
37. Huarng, K., & Yu, H. K. (2005). A type 2 fuzzy time series model for stock index forecasting. *Physica A, 353*, 445–462.
38. Bajestani, N. S., & Zare, A. (2011). Forecasting TAIEX using improved type 2 fuzzy time series. *Expert Systems with Applications, 38*(5), 5816–5821.
39. Zadeh, L. A. (1965). Fuzzy sets. *Information and Control, 8*, 338–353.
40. Wu, D. (2011). A constraint representation theorem for interval type-2 fuzzy sets using convex and normal embedded type-1 fuzzy sets, and its application to centroid computation. In *Proceedings of World Conference on Soft Computing, San Francisco, CA*, May 2011.
41. Karnik, N. N., & Mendel, J. M. (1999). Applications of type-2 fuzzy logic systems to forecasting of time-series. *Information Sciences, 120*, 89–111.
42. Mendel, J. M., Hagras, H., Tan, W. W., Melek, W. W., & Ying, H. (2014). *Introduction to type-2 fuzzy logic control: Theory and applications*. Hoboken, NJ: IEEE-Wiley Press.
43. Chakraborty, S., Konar, A., Ralescu, A., & Pal, N. R. (2015). A fast algorithm to compute precise type-2 centroids for real time control applications. *IEEE Transactions on Cybernetics, 45*(2), 340–353.
44. Mendel, J. M., & Wu, D. (2010). *Perceptual computing: Aiding people in making subjective judgements*. Hoboken, NJ: IEEE-Wiley Press.
45. Edwards, R. P., Magee, J., & Bassetti, W. H. C. (2007). *Technical analysis of stock trends* (9th ed., p. 10). New York, USA: Amacom.
46. TAIEX [Online]. Available: http://www.twse.com.tw/en/products/indices/tsec/taiex.php.
47. Storn, R., & Price, K. G. (1997). Differential Evolution- a simple and efficient heuristic for global optimization over continuous spaces. *Journal of Global Optimization, 11*(4), 341–359.
48. Das, S., & Suganthan, P. N. (2011). Differential evolution: A survey of the state-of-the-art. *IEEE Transactions on Evolutionary Computation, 15*(1), 4–31.
49. Das, S., Abraham, A., Chakraborty, U. K., & Konar, A. (2009). Differential evolution using a neighborhood-based mutation operator. *IEEE Transactions on Evolutionary Computation, 13* (3).
50. Price, K. V., Storn, R. M., & Lampinen, J. A. (Eds.). (2005). *Differential evolution: A practical approach*. New York: Springer.
51. Das, S., et al. (2008). Particle swarm optimization and differential evolution algorithms: Technical analysis, applications and hybridization perspectives. *Studies in Computational Intelligence (SCI), 116*, 1–38.
52. Das, S., & Suganthan, P. N. (2011). Differential evolution: A survey of the state-of-the-art. *IEEE Transactions on Evolutionary Computation, 15*(1), 4–31.
53. Chen, S. M., & Kao, P.-Y. (2013). TAIEX forecasting based on time series, particle swarm optimization techniques and support vector machines. *Information Science, 247*, 62–71.

54. Cai, Q., Zhang, D., Zheng, W., & Leung, S. C. H. (2015). A new fuzzy time series forecasting model combined with ant colony optimization and auto-regression. *Knowledge-Based Systems, 74*, 61–68.
55. Chen, M.-Y. (2014). A high-order fuzzy time series forecasting model for internet stock trading. *Future Generation Computer Systems Elsevier, 37,* 461–467.
56. Askaria, S., Montazerina, N., & Fazel Zarandi, M. H. (2015). A clustering based forecasting algorithm for multivariable fuzzy timeseries using linear combinations of independent variables. *Applied Soft Computing, Elsevier, 35*, 151–160.

# Chapter 4
# Learning Structures in an Economic Time-Series for Forecasting Applications

**Abstract** The chapter introduces a machine learning approach to knowledge acquisition from a time-series by incorporating three fundamental steps. The first step deals with segmentation of the time-series into time-blocks of non-uniform length with distinguishable characteristics from their neighbours. The second step groups structurally similar time-blocks into clusters by an extension of the DBSCAN algorithm to incorporate multilevel hierarchical clustering. The third step involves representation of the time-series by a special type of automaton with no fixed start or end states. The states in the proposed automaton represent (horizontal) partitions of the time-series, while the cluster centres obtained in the second step are used as input symbols to the states. The state-transitions here are attached with two labels: probability of the transition due the input symbol at the current state and the expected time required for the transition. Once an automaton is built, the knowledge acquisition (training) phase is over. During the test phase, the automaton is consulted to predict the most probable sequence of symbols at a given starting state and the approximate time required (within user-defined margin) to reach a user-defined target state with its probability of occurrence. Test phase prediction accuracy being high over 90%, the proposed prediction can be utilized for trading and investment in stock market.

## 4.1 Introduction

A time-series represents a discrete time-valued function describing population growth [1], variation of atmospheric temperature [2] and rainfall [3], economic growth [4] and the like. Prediction of a time-series refers to determining its future values from its current and previous values. Several methods of time-series modeling and prediction are available in the literature [5–11]. Most of these methods are concerned with partitioning the dynamic range of a time-series into fixed size intervals, called partitions, and then determining the set of rules that direct the transitions of the series from a given partition (containing the data point at time $t$) to the next partition (containing the data point at time $t + 1$). The rules obtained



A. Konar and D. Bhattacharya, *Time-Series Prediction and Applications*,
Intelligent Systems Reference Library 127,
DOI 10.1007/978-3-319-54597-4_4

thereby are used to predict the partition containing the data point at time $(t + 1)$ from the given partition, containing the data point at time $t$. Unfortunately, the fluctuation in the time-series [12] is usually governed by external factors, many of which are unknown and unpredictable. Naturally, forecasting of a time-series based on the extracted rules is often not free from errors. This chapter proposes a novel approach for knowledge acquisition from a time-series by incorporating three fundamental steps: segmentation of the series, clustering of the generated segments and representation of the time-series by a special type of automaton [13].

Segmentation is a process by which a time-series is vertically fragmented (sliced) into segments or time-blocks, where each block maintains certain common characteristics different from those of its neighbours. In this chapter, we propose a novel, non-parametric, online segmentation algorithm that slices the time-series into segments based on the local gradient of the transitions between successive data points. Segmentation is performed in three steps. First, transitions of the consecutive data points in a time-series are assigned one of three labels: rise, fall and equality, indicating a positive, a negative and a near-zero slope respectively. In the second step, we construct a dynamic window covering a fixed number (here, 5) of consecutive transitions and assign it a label having the maximum frequency-count of all the three labels within the window. The last step combines successive windows having the same label to form a temporal segment or time-block.

The merits of the proposed segmentation algorithm lies in its (a) simplicity in natural grouping of similar structures of the time-series in a segment, (b) low computational complexity and (c) non-parametric characteristic. The segmented time-blocks of unequal length are next normalized by representing them with a fixed number (here, 10) of equi-spaced data points. Clustering is required here to group similar normalized temporal structures of the segments. Apparently, any traditional clustering algorithm could have been employed to cluster the segments. Since we have no knowledge about the cluster-count, we prefer an algorithm that does not require number of clusters as an input parameter. In addition, we want the highly dense local data points, representative of a specific geometry of segments, to lie in a cluster. Fortunately, the DBSCAN [14] algorithm satisfies both the criteria, and thus is selected as the base clustering algorithm.

The DBSCAN is a parametric algorithm that considers a fixed radius of data-neighbourhood around each data point, and thus may group data points of different density in the same cluster. This undesirable characteristic of DBSCAN has been eliminated by the proposed extension of multi-level (hierarchical) clustering. The said technique clusters data-points of similar density at one level, and passes the un-clustered data-points as outliers for further clustering at subsequent levels. The proposed extension of DBSCAN thus ensures that data-points of low densities even are not disregarded, in case they belong to one or more clusters. The obtained cluster centres are then recorded for subsequent use as input symbols in a specialized automaton, constructed for knowledge representation [15] of the given time-series.

The last step of knowledge acquisition involves construction of the above (stochastic) automaton. The states of the automaton here represent horizontal partitions of a time-series and the state-transitions indicate the transition from one partition to another due to the occurrence of an input symbol (cluster centre describing a temporal segment). Each transition is associated with two labels: (a) the probability of occurrence of the transition and (b) the expected duration of the transition. Unlike traditional automata, the proposed stochastic automaton has no fixed starting or end states as they are determined dynamically by user-defined query, thus signifying its name as dynamic stochastic automaton.

The merit of the above automaton is to autonomously determine the most probable sequence of transitions between given starting and target states (partitions) for a user provided time-limit. Experiments undertaken on the TAIEX [16] close-price economic time-series reveal a very high prediction accuracy over 90% for predicting the most probable sequence of transitions from a given starting state to reach a given target state within a user defined time limit of 90 days. A high prediction accuracy ensures a good knowledge transfer to the stochastic automaton, thereby enabling it to be an effective tool for business forecasting such as trading and investment in stock market.

A performance analysis is undertaken to compare the relative merits of the proposed segmentation and clustering algorithms with traditional ones. We introduce a metric, called *match-score*, to measure the positional similarity of the segment boundaries produced by a segmentation algorithm with hand-crafted ground-truth segment boundaries. Experiments undertaken indicate that the proposed segmentation algorithm outperforms its competitors by a large difference in match-score. For example, the match-score of the proposed algorithm is 76.5% in comparison to the nearest best segmentation algorithm, which has a match-score of 50%. To compare the relative performance of the clustering algorithms, we here use two important metrics, called sum of squared intra-cluster distance (SSID) and Sum of inter-cluster distances (SICD) [17]. Experimental results reveal that the proposed clustering technique outperforms traditional clustering algorithms (k-means [18], fuzzy-c means [19] and DBSCAN) with respect to the above metrics. Further, when compared with the DBSCAN algorithm, we note that the proposed clustering algorithm run on TAIEX time-series data for the period 1990–2000 produces 9 clusters (of diverse geometry) in comparison to only 2 clusters (representative of bullish and bearish patterns [20]) produced by the DBSCAN algorithm.

The chapter is divided into ten main sections. Section 4.2 provides a brief introduction to related existing works on time-series segmentation and clustering. In Sect. 4.3, we provide an overview of the DBSCAN clustering algorithm. Section 4.4 discusses the principles and techniques for the development of a novel time-series segmentation algorithm. In Sect. 4.5, we propose an extension of the DBSCAN clustering algorithm. Section 4.6 is concerned with the construction of a dynamic stochastic automaton. Section 4.7 discusses the computational complexity of the proposed algorithms. Section 4.8 deals with prediction experiments on Taiwan Stock Exchange Index (TAIEX) time-series. Section 4.9 discusses the performance analysis of the proposed time-series segmentation and clustering algorithms. Conclusions are listed is Sect. 4.10.

## 4.2 Related Work

Most of the works on time-series are concerned with its modelling and prediction [5–11]. Yet another field of research on time-series deals with its segmentation. The primary motivation behind segmentation is to reduce the data dimension of the time-series in an efficient manner by selecting certain data points of the time-series, as segment boundaries where significant changes in its slope are observed. Several methods to reduce data points between two consecutive boundaries are available in the literature. A few of these approximation models, which are popularly being used include Piecewise Linear Approximation [21], Piecewise Aggregate Approximation [22] and Singular Value Decomposition [23]. Common metrics used to measure approximation error in segmentation are root mean square error (RMSE) [11] and maximum vertical distance error (MVD) [24]. Both the errors are evaluated by measuring the vertical distances between the regressed/interpolated curve and the time-series within a segment.

Among the well-known time-series segmentation algorithms, the following need special mention. The first one, called the sliding window (SW) algorithm [25], primarily determines the location of the next segment boundary by iteratively widening the current segment, until the approximation error of the segment exceeds a given value. The second algorithm, called the top-down algorithm [26] employs a divide and conquer rule for segmentation of a time-series. It is a recursive algorithm, where each step of recursion attempts to optimally place a segment boundary within a given fragment of a time-series in order to divide it into two parts. The location of the segment boundary is optimal in the sense that the approximation error in the left and right halves, thus formed, is minimal. Unlike top down realization, which employs splitting of the time-series for optimal segmentation, the third algorithm employs a bottom-up [27] approach to merge segments to determine segment boundaries. Starting with each data point as a segment boundary, the algorithm merges segments following a greedy approach until the approximation error of the time-series exceeds a given threshold. The fourth one, called the Sliding Window and Bottom Up (SWAB) algorithm [21] combines the joint benefits of SW and bottom-up techniques to perform online segmentation of a time-series using a buffer. The contiguous blocks of time-series data are placed in the buffer sequentially for online segmentation using the bottom-up algorithm.

Besides the above, there remain a few other interesting algorithms on time-series segmentation. Liu et al. [24], for instance, extend the SW algorithm to reduce its computational overhead. They developed two extensions of SW, referred to as Feasible Space Window (FSW) and the Stepwise Feasible Space Window (SFSW) algorithms. Unlike the SW algorithm, where the approximation error for the entire segment needs to be calculated every time a new data point is added to the segment, both the FSW and SFSW algorithms use a modified segmentation criterion based on the MVD error metric. The segmentation criterion is designed in a way such that the error computation for a newly added point is carried out once, thus improving upon the computational complexity of the SW algorithm. Other noticeable efforts in time-series segmentation include the use of dynamic programming [28], fuzzy clustering [29], least square approximations [30] and evolutionary algorithms [31]. The segmentation algorithm that we develop in this chapter provides a novel non-parametric online approach to natural segmentation of a time-series based on rising, falling or zero slopes of time-blocks in the series.

Apart from segmentation, there exist quite a few other interesting techniques in the field of time-series analysis. Clustering [32] is one of the machine learning [33] based approaches which has found its way into several applications to process time-series data. Clustering is primarily used to detect structural similarity in unlabeled data sets based on various distance metrics like Euclidean distance, Minkowski distance [34] and others. A literature survey of the application of clustering algorithms on time-series data can be found in [34].

As has been mentioned before, for the purpose of identifying clusters of variable densities, we use an extended version of the data density based DBSCAN clustering algorithm. Among the existing extensions of DBSCAN which solve the problem of clustering data with variable densities, the DBSCAN-DLP [35] algorithm requires special mention. In DBSCAN-DLP, the entire data set is pre-processed and ordered into multiple layers based on their density. Each layer is later clustered using the DBSCAN algorithm. Although we employ a similar principle in our approach, the primary difference lies in the application of a greedy recursive technique where, instead of pre-processing the data set, we layer the data set by isolating the maximum density clusters and removing points with lower surrounding density as outliers. The main advantages of such recursive layering are that cluster centres are hierarchically arranged in decreasing order of data density. In addition data points of uniform, density, scattered spatially are clustered at the same level.

## 4.3 DBSCAN Clustering—An Overview

This section outlines the DBSCAN algorithm. DBSCAN is a density-based spatial clustering algorithm that groups points lying in a data-dense region into a cluster and marks points with non-dense surroundings as noise. It requires two input parameters: (i) the radius $\varepsilon$ describing the neighbourhood of a point, and (ii) a threshold $m$ representing the minimum number of points to lie in the neighbourhood of a (randomly or otherwise) selected point. The following terminologies are required to explain the rest of this section.

**Definition 4.3.1** A point $P$ in a given domain of points $D$ is called a **core point**, if there exists a set of points $P' = \{P_1, P_2, P_3, \ldots, P_k\}$, such that the distance between $P_i \in P'$ and $P$ is less than or equal to a pre-assigned small positive number $\varepsilon$ (i.e., the point $P_i$ lies in the **ε-neighbourhood** of $P$) and the number of points $k$ in the set $P'$ is greater than or equal to an empirically selected threshold value $m$ as given in (4.1–4.2).

$$\|P - P_i\| \leq \varepsilon, \quad \forall P_i \in P' \tag{4.1}$$

and

$$k \geq m \tag{4.2}$$

Any point $P_i \in P'$, lying in the **ε-neighbourhood** of $P$ is said to be **directly density-reachable** from the point $P$. **Data-density** of point $P$ here, is defined as $e/\varepsilon$ where $e$ is the number of points in the ε-neighbourhood of $P$.

**Definition 4.3.2** A point $P$ is **density-reachable** from a point $Q$, if there exists a sequence of points $P_1, P_2, \ldots, P_n$ such that $P$ is directly density-reachable from $P_1$, $P_{i+1}$ is directly density-reachable from $P_i$, $\forall i \in \{1, 2, \ldots, n-1\}$ and $Q$ is directly density-reachable from $P_n$. A point $P_k$ is an **outlier**, if it is not density-reachable from any other point.

Pseudo Code 4.3.1: DBSCAN Clustering

***Input:*** *A data set D, of q unlabeled data points,* $\{P_1, P_2, \ldots, P_q\}$ *in k-dimensional space, and user-defined parameters* ε *and m, representing respectively, the neighbourhood radius of a point and minimum number of points around a core point.*
***Output:*** *A q-dimensional vector* $\vec{C} = [c_1, c_2, \ldots, c_q]$, *where* $c_i$ *denotes the cluster number (representative of cluster identity) of the point* $P_i \, \forall i \in \{1, 2, \ldots, q\}$.

*BEGIN*
*Initialize cluster_count* $\leftarrow$ 1, *and declare all points in D as unprocessed;*
*Initialize set N* $\leftarrow \varnothing$;
*FOR each unprocessed point* $P_i$, *in D do*
  *BEGIN*
  *Mark* $P_i$ *as processed;*
  *FOR each point p in D, if p is directly density reachable from* $P_i$, *save p in N;*
  *IF* ($|N| < m$) *THEN ignore* $P_i$ *as noise;*
  *ELSE do*
  *BEGIN*
    *Mark* $P_i$ *as a core point and assign cluster_count as cluster number* $c_i$ *to* $P_i$;
    *FOR each unprocessed point* $P_j$ *in N, do*
    *BEGIN*
      *Mark* $P_j$ *as processed; Initialize set M* $\leftarrow \varnothing$;
      *FOR each point s in D, if s is directly density reachable from* $P_j$, *save s in M;*
      *IF* ($|M| \geq m$) *THEN* $N \leftarrow N \cup M$;
      $N \leftarrow N - \{P_j\}$;
      *Assign cluster_count as cluster number* $c_j$ *to* $P_j$;
    *END FOR;*
  *END IF;*
*Increment cluster_count by one; Reinitialize N* $\leftarrow \varnothing$;
*END FOR;*
*END.*

**Definition 4.3.3** Two points $P_i$ and $P_j$ are **density-connected,** if there exists a point $P$ such that both $P_i$ and $P_j$ are density-reachable from $P$.

A pseudo code for the DBSCAN algorithm is given in below. In this algorithm, we call a point $P_i$ processed, when it is selected to examine the points in its $\varepsilon$-neighbourhood. Any point not processed so far is called unprocessed.

The algorithm includes three main steps.

1. It begins by selecting an arbitrary unprocessed point $P_i$ from the given data set $D$ and checks if the point is a core point. If not, $P_i$ is ignored as noise. However, if it is a core point, the algorithm identifies all points in the $\varepsilon$-neighbourhood of $P_i$ and stores them in a set $N$. This set is used to store points which are yet to be processed for the creation of an individual cluster.
2. Next, the algorithm processes each point $P_j$ in the set $N$ by finding points lying in the $\varepsilon$-neighbourhood of $P_j$ (storing them in a separate set $M$) and including them in $N$, (i.e., $N \leftarrow N \cup M$). The inclusion of new unprocessed points in $N$, with progressively increasing distance from the initial core point leads to the growth of the cluster boundary. When all points in $N$ are processed, a cluster is formed and a new core point is searched.
3. The algorithm ends when unprocessed core points can no longer be detected.

It should be noted that a cluster always has the following two properties.

**Property 4.3.1** Any two points $P_i$ and $P_j$, lying in a cluster are mutually density-connected.

**Property 4.3.2** If $P_i$ is a point which is density-reachable from point $P_j$ and $P_j \in C$, where $C$ is a cluster, then $P_i \in C$.

## 4.4 Slope-Sensitive Natural Segmentation

In this section, we propose a novel algorithm for Slope-Sensitive Natural Segmentation (SSNS) of a time-series into time-blocks of unequal lengths. The principles adopted in the algorithm follow from the natural labelling of ridges in a mountain as rising/falling and plateau (zero slope). The algorithm intuitively attempts to determine the transitions of consecutive pairs of data points in a time-series into one of three distinct categories (labels): rise (R), fall (F) and equality (E), based on the measure of an approximate slope of the straight line joining the points. The measure of slope is determined by dividing the dynamic range of the time-series into a fixed number of horizontal partitions, and then by examining their relative positions along the y-axis of the time-series. A rise (fall) refers to a relative rise (fall) in the position of the partitions containing the consecutive data points over time, whereas an equality indicates occupancy of both the data points in the same partition. After the transitions are labelled, we group five consecutive transitions, called a window, and assign the window a label having the highest count (frequency) of occurrence in the window. In case there is a tie of two labels, we break it by additional constraints. A segment refers to a sequence of consecutive windows having the same label. Segmentation deals with vertically slicing a time-series at the centre of a window, when the next window has a different label with respect to the current window.

### 4.4.1 Definitions

**Definition 4.4.1** A **time-series** is a discrete sequence of samples of a measurable entity, such as temperature, atmospheric pressure, and population growth, taken over a finite interval of time. For an entity $x$, varying over time $t$ in $[t_{min}, t_{max}]$, the time-series of length $n$ can be expressed as a vector $\vec{X} = [x(kT)] = [x_k]$, 0 where $T$ is the sampling interval and $k$ is an integer in $[1, n]$.

**Definition 4.4.2** Given a time-series $\vec{X}$, let $X_{max}$ and $X_{min}$ respectively denote the maximum and minimum sample values of the series. Partitioning the time-series here refers to dividing the range $Z = [X_{min}, X_{max}]$ into $k$ non-overlapping contiguous blocks $Z_1, Z_2, \ldots, Z_k$, called **partitions**, such that the following two conditions jointly hold:

$$Z_p \bigcap Z_q = \varnothing, \quad \forall p, \forall q \in \{1, 2, \ldots, k\}, p \neq q \tag{4.3}$$

$$\bigcup_{i=1}^{k} Z_i = Z, \quad \text{for integer } i \tag{4.4}$$

*Example 4.4.1* Let $\vec{X} = [2, 5, 4, 10, 3]$ be a time-series. According to Definition 4.4.2, $X_{min} = 2$ and $X_{min} = 10$. We divide the range $[2, 10]$ into four partitions $Z_1 = [2, 4)$, $Z_2 = [4, 6)$, $Z_3 = [6, 8)$ and $Z_4 = [8, 10]$. It should be noted that the bounds of $Z_2$ upper bound of $Z_2$ and lower bound of $Z_4$ are not present in $\vec{X}$. However, such partitioning is allowed by Definition 4.4.2.

**Definition 4.4.3** Let $Z_p = \left[Z_p^-, Z_p^+\right]$ and $Z_q = \left[Z_q^-, Z_q^+\right]$ be two partitions in a time-series $\vec{X}$ and $(x_i, x_{i+1})$ be two successive data points in the series, such that $x_i \in Z_p$ and $x_{i+1} \in Z_q$. Then the **transition** of $x_i$ to $x_{i+1}$, denoted by $Tr(x_i, x_{i+1})$, is assigned one of three possible linguistic labels: Rise (R), Fall (F) and Equality (E) using (4.5).

$$Tr(x_i, x_{i+1}) = \begin{cases} R, & \text{if } Z_p^+ \leq Z_q^- \\ F, & \text{if } Z_p^- \leq Z_q^+ \\ E, & \text{if } Z_p^+ = Z_q^+ \text{ and } Z_p^- = Z_q^- \end{cases} \tag{4.5}$$

*Example 4.4.2* Considering the time-series $\vec{X}$ given in Example 4.4.1, $x_1 = 2$ and $x_2 = 5$. Thus, $x_1 \in Z_p = [2, 4)$ and $x_2 \in Z_q = [4, 6)$. Since, $Z_q^- \geq Z_p^+$, by Definition 4.4.3, $Tr(x_1, x_2) = R$. □

**Definition 4.4.4** For a time-series $\vec{X} = [x_1, x_2, \ldots, x_n]$, a string of the form $\vec{S} = [S_1, S_2, \ldots, S_{n-1}]$ where $S_i = Tr(x_i, x_{i+1})$, for any integer $i \in [1, n-1]$, is called a **ternary string of transitions (T-SOT)**.

*Example 4.4.3* For the time-series $\vec{X}$ of Example 4.4.1, $x_1 = 2$, $x_2 = 5$, $x_3 = 4$, $x_4 = 10$ and $x_5 = 3$. It can be verified using $Z_1$, $Z_2$, $Z_3$ and $Z_4$ that $Tr(x_1, x_2) = R$; $Tr(x_2, x_3) = E$; $Tr(x_3, x_4) = R$ and $Tr(x_4, x_5) = F$. Thus, the ternary string of transitions $\vec{S}$ for $\vec{X}$ is given by $\vec{S} = [R, E, R, F]$. □

**Definition 4.4.5** Let $\vec{S} = [S_1, S_2, S_3, \ldots, S_{n-1}]$ be a T-SOT constructed from the time-series $\vec{X} = [x_1, x_2, \ldots, x_n]$. For each character $S_i$, a string of length $L \geq 3$ characters selected from $\vec{S}$ centred around $S_i$ is referred to as **window** $\vec{W}_i$. For instance, a window $\vec{W}_i$ of length $L$ characters is given in (4.6).

$$\vec{W}_i = \left[S_{i-\lfloor L/2 \rfloor}, S_{i-\lfloor L/2 \rfloor+1}, \ldots, S_i, \ldots, S_{i+\lfloor L/2 \rfloor-1}, S_{i+\lfloor L/2 \rfloor}\right] \tag{4.6}$$

where $i \in \{\lfloor L/2 \rfloor + 1, \lfloor L/2 \rfloor + 2, \ldots, n - \lfloor L/2 \rfloor - 1\}$. Here onwards, we use a window of five linguistic characters, of the form $\vec{W}_i = [S_{i-2}, S_{i-1}, S_i, S_{i+1}, S_{i+2}]$, for the development of the SSNS algorithm.

*Example 4.4.4* Let $\vec{S} = [R, F, E, E, R, R] = [S_1, S_2, S_3, S_4, S_5, S_6]$ be a T-SOT. Here, $n - 1 = 6$ and hence, $n - 3 = 4$. So, we have two valid windows of length, $L = 5$ characters: $\vec{W}_3 = [R, F, E, E, R]$ and $\vec{W}_4 = [F, E, E, R, R]$ for $i \in \{3, 4\}$. □

**Note 4.4.1** It is important to note that windows containing a fixed length of $r$ linguistic symbols refer to a time-series of length $r + 1$ data points. □

**Definition 4.4.6** Let $\vec{X} = [x_1, x_2, \ldots, x_n]$ be a time-series and $\vec{S} = [S_1, S_2, S_3, \ldots, S_{n-1}]$ be the T-SOT constructed from $\vec{X}$. Let $\vec{W}_i = [S_{i-2}, S_{i-1}, S_i, S_{i+1}, S_{i+2}]$ for $3 \leq i \leq n-3$ be the $i$th window of the time-series. Let $f_x(\vec{W}_i)$ denote the frequency count (number of occurrences) of the linguistic character $x \in \{R, F, E\}$ in $\vec{W}_i$. The window $\vec{W}_i$, is assigned a **label** $L(\vec{W}_i)$ based on the following policy.

$$L(\vec{W}_i) = R, \text{ if } f_R(\vec{W}_i) > f_x(\vec{W}_i), \quad x \in \{F, E\} \tag{4.7}$$

$$= E, \text{ if } f_E(\vec{W}_i) > f_x(\vec{W}_i), \quad x \in \{R, F\} \tag{4.8}$$

If there exists three different linguistic characters: $p, q, r \in \{R, F, E\}$, such that $f_p(\vec{W}_i) = f_q(\vec{W}_i) > f_r(\vec{W}_i)$, then

$$L(\vec{W}_i) = p, \quad \textit{if}\ L(\vec{W}_{i-1}) = p \tag{4.9}$$

$$= q, \quad \textit{if}\ L(\vec{W}_{i-1}) = q \tag{4.10}$$

$$= p \text{ or } q \text{ arbitrarily otherwise.} \tag{4.11}$$

**Note 4.4.2** In Definition 4.4.6, we assign a label $x \in \{R, F, E\}$ to a window $\vec{W}_i$, if the frequency count of $x$ is the highest in $\vec{W}_i$. Furthermore, if frequency count of any two of the three labels are equal and higher than the rest, and if any one of these labels has already been assigned to $\vec{W}_{i-1}$, then we assign the same label to $\vec{W}_i$. In case, the former condition holds but the latter condition fails, we can arbitrarily assign any one of the labels with the highest frequency count, to $\vec{W}_i$. □

*Example 4.4.5* For a given time-series $\vec{X}$, let $\vec{S} = [R, F, R, E, R, R, E, E, F, E]$ be a T-SOT. The windows in the above series are $\vec{W}_3 = [R, F, R, E, R]$, $\vec{W}_4 = [F, R, E, R, R]$, $\vec{W}_5 = [R, E, R, R, E]$, $\vec{W}_6 = [E, R, R, E, E]$, $\vec{W}_7 = [R, R, E, E, F]$, and $\vec{W}_8 = [R, E, E, F, E]$. Since, $f_R(\vec{W}_3) = 3 > f_E(\vec{W}_3) = f_F(\vec{W}_3) = 1$, hence, $L(\vec{W}_3) = R$.

Similarly, $L(\vec{W}_4) = R$ and $L(\vec{W}_5) = R$. In $\vec{W}_6, f_E(\vec{W}_6) = 3 > f_R(\vec{W}_6) = 2 > f_F(\vec{W}_6) = 0$. Hence, $L(\vec{W}_6) = E$. In $\vec{W}_7, f_E(\vec{W}_7) = 2 = f_R(\vec{W}_7) > f_F(\vec{W}_7) = 1$. Since, $L(\vec{W}_6) = E$, following definition 4.4.6, $L(\vec{W}_7) = E$.

Again, as $f_E(\vec{W}_8) > f_R(\vec{W}_8) = f_F(\vec{W}_8)$, $L(\vec{W}_8) = E$ follows. □

**Definition 4.4.7** A **structure** is a collection of one or more contiguous windows which have been assigned the same label.

*Example 4.4.6* In Example 4.4.5, $L(\vec{W}_3) = L(\vec{W}_4) = L(\vec{W}_5) = R$. So, $\vec{W}_3\vec{W}_4\vec{W}_5$ together forms a structure having the same label $R$, as those of the involved windows. The label of the $i$th structure is denoted by $L(struct_i)$. □

**Definition 4.4.8** For two consecutive structures $struct_i$ and $struct_{i+1}$, if $L(struct_i) \neq L(struct_{i+1})$, a segment boundary exists at the centre point $S_j$ of the last window $\vec{W}_j$ (say) of $struct_i$. Since, $S_j$ denotes the transition between time-series data points $x_j$ and $x_{j+1}$, the segment boundary is placed in between these two data points.

*Example 4.4.7* In Examples 4.4.5 and 4.4.6, windows $\vec{W}_3$ to $\vec{W}_5$ in conjunction, form a structure having the label $R$. Again, $\vec{W}_6$ to $\vec{W}_8$ form a structure having the label $E$. Since, the labels of these two consecutive structures are not equal, a segment boundary appears at the centre point $S_5$, of the last window $\vec{W}_5$ in the first structure. As $S_5$ denotes the transition between time-series data points $x_5$ and $x_6$, the segment boundary is placed in between these two data-points. □

### 4.4.2 The SSNS Algorithm

The SSNS algorithm includes four steps. In the first step, we partition the time-series horizontally into $k$ intervals of uniform width $w$ as given in (4.12)

$$w = \frac{1}{n-1}\sum_{i=1}^{n-1} |x_{i+1} - x_i| \tag{4.12}$$

where $x_i$ and $x_{i+1}$ for $i = 1$ to $n - 1$ are consecutive points in the time-series. In case a partition is empty, we merge it with its immediate lower partition. This ensures that no partition is empty and thus, helps in capturing small transitions in the time-series. It may be noted that the lower-most and upper-most partitions being at the boundaries of the dynamic range (=$k \times w$) of the time-series, includes at least one point. Next three steps of segmentation are transition labelling, window labelling and segment boundary determination. They directly follow from the definitions introduced above and are point-wise included in Pseudo Code 4.1.

Pseudo Code 4.1: SSNS Segmentation

***Input:*** *A time-series* $\vec{X} = [x_1, x_2, \ldots, x_n]$, *containing n real valued data points.*
***Output:*** *A vector* $\vec{I} = [i_1, i_2, \ldots, i_m]$, *where* $i_j$ *is the index or the time instant of the* jth *segment boundary.*

***Step 1. Partitioning:*** *Evaluate partition width, w following* (4.12). *Partition the entire range* $[min(\vec{X}), max(\vec{X})]$, *of the time-series* $\vec{X}$ *into k partitions, where* $k = (max(\vec{X}) - min(\vec{X}))/w$ ($min(\vec{X})$ *and* $max(\vec{X})$ *respectively denote minimum and maximum elements of the time-series). Merge partitions containing no data points with their immediate lower partition.*
***Step 2. Transition labelling:*** *For each two consecutive data points* $x_i$ *and* $x_{i+1}$, $1 \leq i \leq (n-1)$, *determine transition between* $x_i$ *and* $x_{i+1}$ *following* (4.5). *Store each computed transition in a T-SOT* $\vec{S} = [S_1, S_2, \ldots, S_{n-1}]$ *where* $S_i = Tr(x_i, x_{i+1}), \forall i \in \{1, 2, \ldots, n-1\}$.
***Step 3. Window labelling:*** *For each group of five consecutive symbols in the T-SOT* $\vec{S}$, *obtain the windows* $\vec{W}_i$ *following* (4.4) *for* $i = 3$ *to* $(n-3)$. *Assign the label* $L(\vec{W}_i)$ *to the window* $\vec{W}_i$, *following* (4.7–4.11).
***Step 4. Segmentation:*** *For two successive windows,* $\vec{W}_{i-1}$ *and* $\vec{W}_i$, *if* $L(\vec{W}_{i-1}) \neq L(\vec{W}_i)$, *then place a segment boundary at the centre point of* $\vec{W}_{i-1}$ *(i.e., at symbol* $S_{i-1}$*) and correspondingly at the centre of* $x_{i-1}$ *and* $x_i$.

A trace of the SSNS segmentation algorithm for the input time-series $\vec{X} = [2, 4, 3, 4, 6, 5, 8, 9, 10, 8, 9, 7, 6, 5]$ has been shown in Fig. 4.1 for illustration.

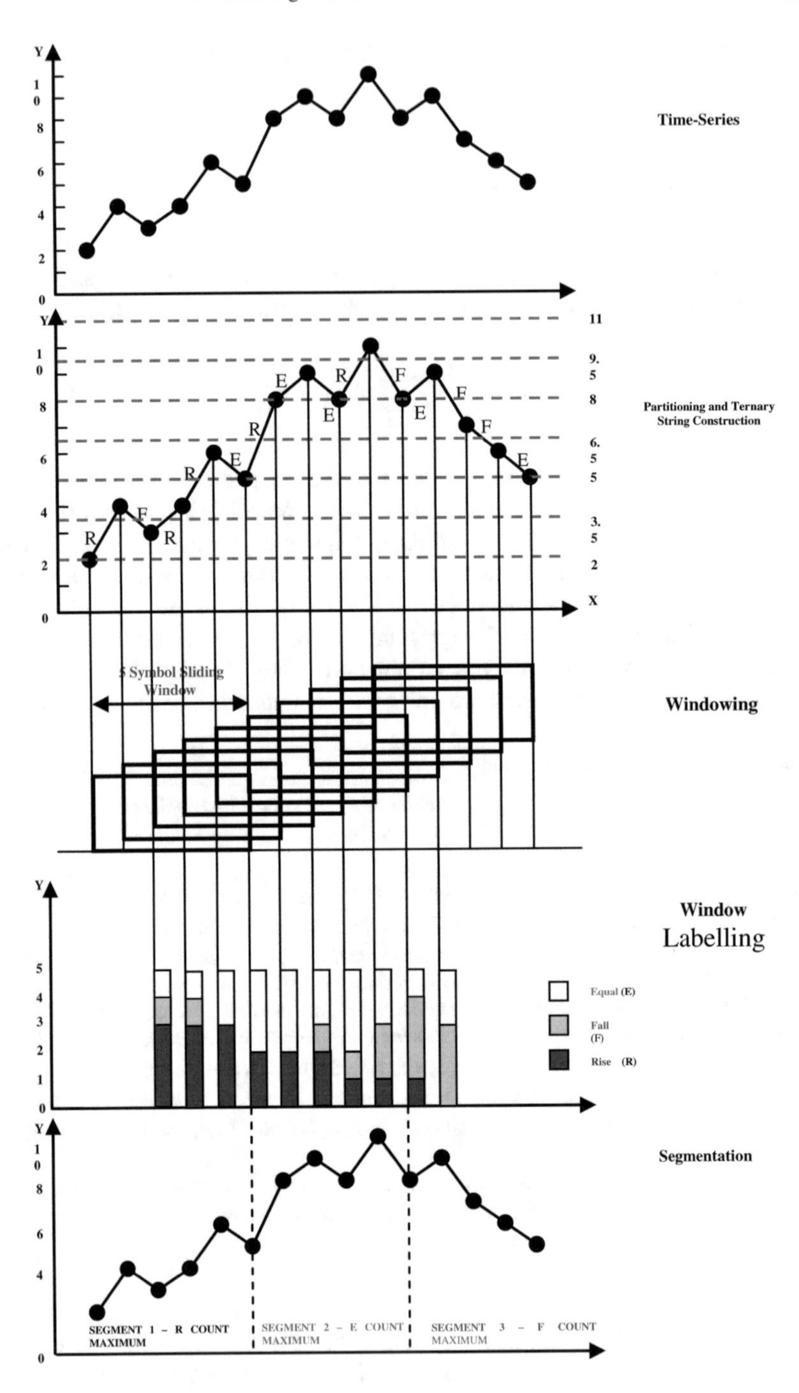


**Fig. 4.1** Trace of SSNS algorithm

## 4.5 Multi-level Clustering of Segmented Time-Blocks

After a time-series is segmented into time-blocks, we attempt to group them into clusters based on their structural similarity. Since the segmented time-blocks are of non-uniform length, clustering is preceded by a pre-processing step, where the time-blocks are re-organized as a pattern of uniform length. The uniform length patterns having varied ranges need to be normalized by additional transformation (Z-score standardization [36]).

After pre-processing, we go for clustering the pre-processed time-blocks by using an extension of DBSCAN clustering, where the motivation is to cluster equally dense regions at the same level, and pass the less dense data points to the next level of clustering in a recursive manner until the data points are less than a prescribed threshold.

### 4.5.1 *Pre-processing of Temporal Segments*

Pre-processing is a two-step process. The first step includes transforming the variable length segmented time-blocks into vectors of uniform (here, 10) length. This is undertaken by the following three sub-steps.

1. Join each pair of consecutive points in a segmented time-block by a straight line, thereby generating a piecewise linear curve.
2. Divide the entire duration of the segment into ten equal parts and mark the corresponding time-points.
3. Determine the ordinates for the marked points on the time-axis of the curve obtained in step 1.

The second step of pre-processing is required to normalize the range of the time-blocks. Let there be $l$ temporal segments in the time-series. Then a matrix $M$ of $(10 \times l)$ elements can be used to store the representation vectors of all the temporal segments, where the $i$th column in $M$ corresponds to the $i$th temporal segment. In order to scale the temporal segments we use Z-score standardization as shown in (4.13).

$$M_{i,j} = \frac{M_{i,j} - mean(M,j)}{std(M,j)} \tag{4.13}$$

where $mean(M,j)$ and $std(M,j)$ are the mean and the standard deviation of the $j$th column of matrix $M$ respectively. Z-score standardization scales the temporal segments to have a zero mean and a unit variance.

### 4.5.2 Principles of Multi-level DBSCAN Clustering

The proposed multi-level DBSCAN clustering is a recursive extension of the DBSCAN algorithm, where at each level of recursion we cluster the data points of highest available density, and pass on the outliers (lower density data points) at that level for further clustering at the next level. This is illustrated in Fig. 4.2. To efficiently undertake the operations at a given level of recursion, we need two additional computations. First, we require evaluating the ε parameter at each level, and next we need to construct a point information table (PIT) at that level.

The PIT stores the number of points within an ε-radius of all possible points, if the number of neighbourhood points around that point exceeds a given threshold. It can be organized by different approaches. Given a set of $l$ points of the form $P = \{p_1, p_2, \ldots, p_l\}$, one simple approach to realize this (PIT) is to allocate an array of pointers, where the $i$th element points to a linear linked list containing the count $N_i$ and indices of the points lying in the ε-neighbourhood of the $i$th point $p_i$. The array of pointers (Fig. 4.3) only point to the linked list of $(l - m)$ points, that satisfy the necessary criterion that the number of data points in the ε-neighbourhood of a point exceeds a pre-defined threshold. In other words, out of $l$ points only $(l - m)$ points satisfy the necessary criterion, and remaining $m$ points do not satisfy it and thus are treated as outliers at that level.

The primary purpose of the PIT is to compute the outliers in each density based stratum. A point $p_i$ is defined as an outlier if the number of points $N_i$ in the ε-neighbourhood of $p_i$ is less than the average of the number of neighbourhood points of all $l$ points in the given data set. This is given in (4.14).

$$\left.\begin{aligned} outlier(p_i) &= true, \quad \text{if } N_i < \left(\frac{1}{l}\right)\sum_{j=1}^{l} N_j \\ &= false, \quad \text{otherwise.} \end{aligned}\right\} \tag{4.14}$$

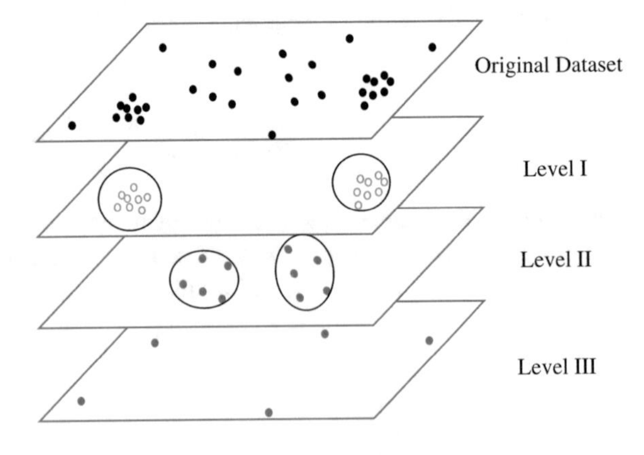


**Fig. 4.2** Multilevel density-based clustering: points of higher density are clustered at higher levels, transferring lower density points for clustering at lower levels

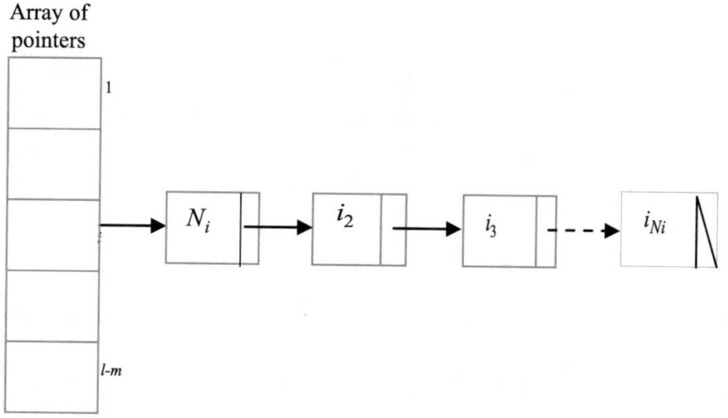


$N_i$= Number of points in the $\varepsilon$ -neighbourhood of $P_i$ where $P_i$ is the $i^{th}$ point in the data set.

**[$i_1$,$i_2$, ..., $i_{Ni}$]= indices of points which lie in the $\varepsilon$ -neighbourhood of $P_i$.**

**Fig. 4.3** Structure of the point information table

Furthermore, the PIT is scanned to identify the point having highest neighbourhood density as the seed for generating a new cluster. This ensures that cluster centres are hierarchically arranged in decreasing order of density.

*Determining the* ε *parameter for each layer:* Given a set of $l$ 10-dimensional points representing Z-score standardized time-blocks, we construct an $l \times l$ distance matrix $D$, whose $(i, j)$th element $d_{i,j}$ represents the Euclidean distance between the points $p_i$ and $p_j$. Now, to compute the ε parameter for the first layer, we sort the distances $d_{i,j}$ in ascending order of their magnitude and take the mean of the first $k$ distances in the sorted list, where $k$ is an empirically chosen parameter, typically taken as 10% of the members in the list. The process of ε parameter computation is repeated in the same manner for each layer of data points.

The proposed multilevel DBSCAN algorithm includes the following three steps in each layer of recursion. From a given set of data points in a layer, the algorithm evaluates the ε-parameter by the procedure introduced above.

The PIT for the given level is constructed to eliminate outliers at that level and also to select the points with highest neighbourhood density as core points for clustering at that level using the DBSCAN algorithm.

The outliers at a given level are defined as the input data set for clustering at the next level. The following property indicates that no additional outliers can be obtained after clustering at a given level.

**Property 4.5.1** *The clustering at each layer of similar density data points does not produce any outliers other than those dropped from the PIT constructed for that layer.*

*Proof* We prove the property by the method of contradiction. Let us assume that after clustering at a given layer, there exist additional outliers excluding those dropped from the PIT for that layer. Outliers being points with less than threshold number of ε-neighbourhood points, are dropped from the PIT. Thus, the points considered for clustering at the layer do not include any point containing less than the threshold number of ε-neighbourhood points. Hence, all points in a given layer are clustered. This contradicts the assumption that there exist additional outliers at a given layer of clustering, and hence its contradiction is true. □

Pseudo Code 4.5.1: Multilevel DBSCAN Clustering

***Input:*** *An array of l 10-dimensional points,* $P = [p_1, p_2 \ldots, p_l]$ *and an empirically chosen threshold value T.*

***Output:*** *A vector* $\vec{C} = [c_1, c_2, \ldots, c_m]$ *of m cluster centres of 10-dimensions each.*

```
BEGIN MAIN
D ← compute_distance_matrix(P);/*following section 4.5.2*/
C ← multilevel_DBSCAN(P, D, T);
END MAIN
```

```
Procedure multilevel_DBSCAN(P, D, T):
BEGIN
q ← num_points(P);/*compute point count in array P */
  IF (q < T) THEN EXIT;/*Stopping criterion of recursion */
ε ← compute_ε(D);/*compute ε following section 4.5.2*/
  Initialize PIT as an array of q pointers all set to NULL;
  Initialize P' as an empty array;/* P' will store outliers */
  FOR each point p_i in P do/*Construction of PIT */
  BEGIN
    Store the index j of all points p_j lying in the ε-neighbourhood of p_i in the ith row of the PIT;
  END FOR;
  FOR each point p_i in P do/*Removal of outliers*/
  BEGIN
    /* outlier(p_i) is computed following equation 4.5.2 */
    IF outlier(p_i) = true, mark p_i as outlier and store it in P';
    ELSE mark p_i as non-outlier;
  END FOR;
  Cluster the non-outlier points using modified-DBSCAN and store corresponding cluster
centres in C⃗;
  Call multilevel_DBSCAN(P', D, T);
END.
```

### 4.5.3 The Multi-level DBSCAN Clustering Algorithm

The modified-DBSCAN algorithm is different from traditional DBSCAN by a single issue only. The modification lies in core point selection. In traditional DBSCAN, core points are selected randomly, whereas in the modified-DBSCAN

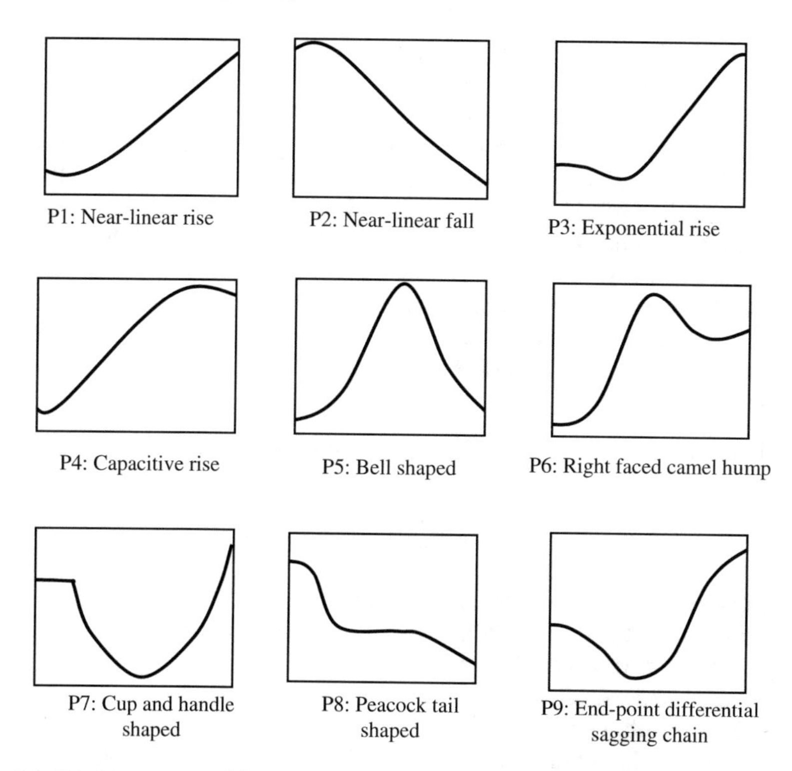


**Fig. 4.4** Primitive patterns (cluster centres) extracted from the TAIEX Time-series with their proposed names based on similarity with real world objects

the point with the highest neighbourhood density in the PIT is selected as the core point for generation of an individual cluster. The modification in core point selection is advantageous for layered clustering as it generates cluster centres in decreasing order of density without requiring additional time-complexity (as analyzed in Sect. 4.7).

The proposed clustering algorithm has successfully been applied to TAIEX close price time-series for the period 1990–2000 with resulting 9 (nine) cluster centres of diverse geometry as illustrated in Fig. 4.4. These cluster centres are used in the next stage for knowledge representation using an automaton.

## 4.6 Knowledge Representation Using Dynamic Stochastic Automaton

This section provides a general method to represent the knowledge acquired from a time-series in the form of an automaton. Here, we use the partitions of the time-series as states in the automaton. The transition from one state to another is triggered by an input symbol, representing one of $m$(=9 for TAIEX) possible cluster centres, describing a fragment of the time-series. We attach three labels with each arc of the automaton describing state transitions. They are (i) the input symbol, (ii) probability of the transition due to the input symbol and (iii) the expected duration of the transition, all in order. We write them as $a/x$, $y$ where '$a$' is the input symbol, '$x$' is the probability of transition and '$y$' is the expected duration.

The automaton outlined above helps in determining the expected time to reach a target state (partition) from a given starting state following the most probable sequence of consecutive transitions. Alternatively, it can also be used to ascertain if the probability to reach a given target state crosses a given threshold within a user-provided time-limit. The formal definition of the automaton is given below for convenience.

**Definition 4.6.1** A **Dynamic Stochastic Automaton (DSA)** is defined by a 5-tuple given by

$$F = (Q, I, V, \delta, \delta_t) \tag{4.15}$$

where, $Q$ = Non-empty finite set of states (partitions)

$I$ = Non-empty finite set of input symbols, representative of primitive patterns (cluster centres) obtained in Sect. 4.5

$V$ = [0,1] is the valuation space

$\delta$ = Probabilistic transition function $\delta : Q \times I \times Q \to V$

$\delta_t$ = Temporal transition function $\delta : Q \times I \times Q \to \mathbf{R}$, where $\mathbf{R}$ is the set of real numbers representing time required for the transition. It is required that for any given state $q$ and any given input symbol $p$, the sum of the probability of transitions to all possible next states $q' \in Q$ is given by

$$\sum_{q' \in Q} \delta(q, p, q') = 1 \tag{4.16}$$

We now define an extended probabilistic transition function $\delta' : Q \times I^* \times Q \to V$ that can recognize a sequence of consecutive input symbols $I^* = <p_1, p_2, \ldots, p_n>$ by the following notation

$$\delta'(q, p_1 p_2 \ldots p_n, q') = \sum_{q_1 \ldots q_{n-1} \in Q} [\delta(q, p_1, q_1)\delta(q_1, p_2, q_2)\ldots\delta(q_{n-1}, p_n, q')] \tag{4.17}$$

$\delta_t' : Q \times I^* \times Q \to \mathbf{R}$ for the sequence $I^* = <p_1, p_2, \ldots, p_n>$ to represent the minimum time required for the transitions to traverse the entire sequence $I^*$ of input symbols. This is given in (4.18).

$$\delta_t'(q, p_1 p_2 \ldots p_n, q') = \min_{q_1 \ldots q_{n-1} \in Q} \{\delta_t(q, p_1, q_1) + \delta_t(q_1, p_2, q_2) + \cdots + \delta_t(q_{n-1}, p_n, q')\} \tag{4.18}$$

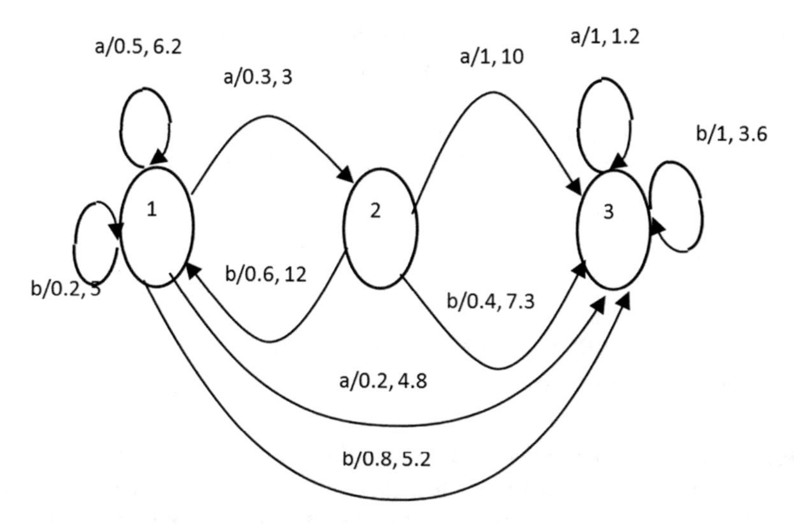


**Fig. 4.5** A DSA to illustrate its parameters

**Note 4.6.1** It is to be noted that the proposed DSA differs from traditional stochastic automata [13] by the following counts. First, unlike traditional stochastic automata, the DSA has no fixed starting states and terminal states, as they are user-defined and thus dynamically chosen, justifying the name, dynamic stochastic automaton. Second, there is an additional term $\delta_t$ describing the time required for the transition to take place on occurrence of an input symbol $p \in I$ to a given state $s \in Q$. □

*Example 4.6.1:* This example illustrates the notations used in (4.15), (4.17) and (4.18) using the automaton shown in Fig. 4.5.

Here, $Q = \{1, 2, 3\}$, $I = \{a, b\}$,

$$\begin{aligned}
&\delta(1,a,1) = 0.5,\ \delta(1,a,2) = 0.3,\ \delta(1,a,3) = 0.2,\ \delta(2,a,3) = 1,\\
&\delta(3,a,3) = 1,\ \delta(1,b,1) = 0.2,\ \delta(1,b,3) = 0.8,\ \delta(2,b,1) = 0.6,\\
&\delta(2,b,3) = 0.4,\ \delta(3,b,3) = 1,\\
&\delta_t(1,a,1) = 6.2,\ \delta_t(1,a,2) = 3,\ \delta_t(1,a,3) = 4.8,\ \delta_t(2,a,3) = 10,\\
&\delta_t(3,a,3) = 1.2,\ \delta_t(1,b,1) = 5,\ \delta_t(1,b,3) = 5.2,\ \delta_t(2,b,1) = 12,\\
&\delta_t(2,b,3) = 7.3,\ \delta_t(3,b,3) = 3.6.
\end{aligned}$$

We now illustrate the evaluation of the probability of occurrence of the sequence of input symbols $I^* = <a, a>$ at starting state 1 and terminal state 3.

$$\begin{aligned}
\delta'(1,aa,3) &= \delta(1,a,1)\delta(1,a,3) + \delta(1,a,2)\delta(2,a,3) + \delta(1,a,3)\delta(3,a,3)\\
&= (0.5 \times 0.2) + (0.3 \times 1) + (0.2 \times 1) = (0.1 + 0.3 + 0.2) = 0.6
\end{aligned}$$

Further, we can evaluate the minimum transition time for the paths starting at state 1 and terminating at state 3 following the pattern sequence $I^* = <a, a>$ by

$$\begin{aligned}\delta_t'(1, aa, 3) &= \min\{\delta_t(1, a, 1) + \delta_t(1, a, 3), \delta_t(1, a, 2) + \delta_t(2, a, 3), \delta_t(1, a, 3) + \delta_t(3, a, 3)\} \\ &= \min\{6.2 + 4.8,\ 3 + 10,\ 4.8 + 1.2\} = \min\{11,\ 13,\ 6\} = 6\end{aligned}$$

### 4.6.1 Construction of Dynamic Stochastic Automaton (DSA)

Construction of an automaton requires partitioning the dynamic range of the time-series into fewer partitions, rather than large number of partitions as undertaken during the segmentation phase. Apparently this looks a little tricky. However, the re-partitioning scheme has a fundamental basis to ensure that the partitions, when transformed to states in the DSA, will be utilized in most cases if they are relatively fewer. In case the partitions obtained during segmentation are used here, most of the states in the automaton remain non-utilized, as the vertical span of the temporal patterns obtained in clustering is large enough to cover several partitions obtained in the segmentation phase. After careful experiments with the TAIEX time-series, it is observed that the temporal segments obtained by clustering can cover up to 3–4 partitions if the dynamic range is equally partitioned into 7–10 partitions. In the illustrative time-series of Fig. 4.6a, we have partitioned it into 5 partitions.

The computation of the probability of a transition $\delta(s_i, a, s_j)$ from state $s_i$ to state $s_j$ due to input symbol $a$ where $s_i, s_j \in Q$ and $a \in I$, is obtained as

$$\delta(s_i, a, s_j) = \frac{freq\_cnt(s_i, a, s_j)}{\sum_{s_k \in Q} freq\_cnt(s_i, a, s_k)} \tag{4.19}$$

where $freq\_cnt(s_i, a, s_j)$ denotes the number of transitions from partition $s_i$ to partition $s_j$ due to the occurrence of a temporal pattern belonging to cluster $a$ (that acts as an input symbol).

The computation of $\delta_t(s_i, a, s_j)$ is obtained here by measuring the duration of all possible transitions in the time-series with starting state $s_i$ and terminating state $s_j$ due to the occurrence of a temporal pattern belonging to cluster $a$. Let the possible temporal segments falling in cluster $a$ that appear at partition $s_i$ for transition to partition $s_j$ be $a_1, a_2, \ldots, a_n$. Then, $\delta_t(s_i, a, s_j)$ is obtained as

$$\delta(s_i, a, s_j) = \frac{\sum_{k=1}^{n} \tau(s_i, a, s_j)}{n} \tag{4.20}$$

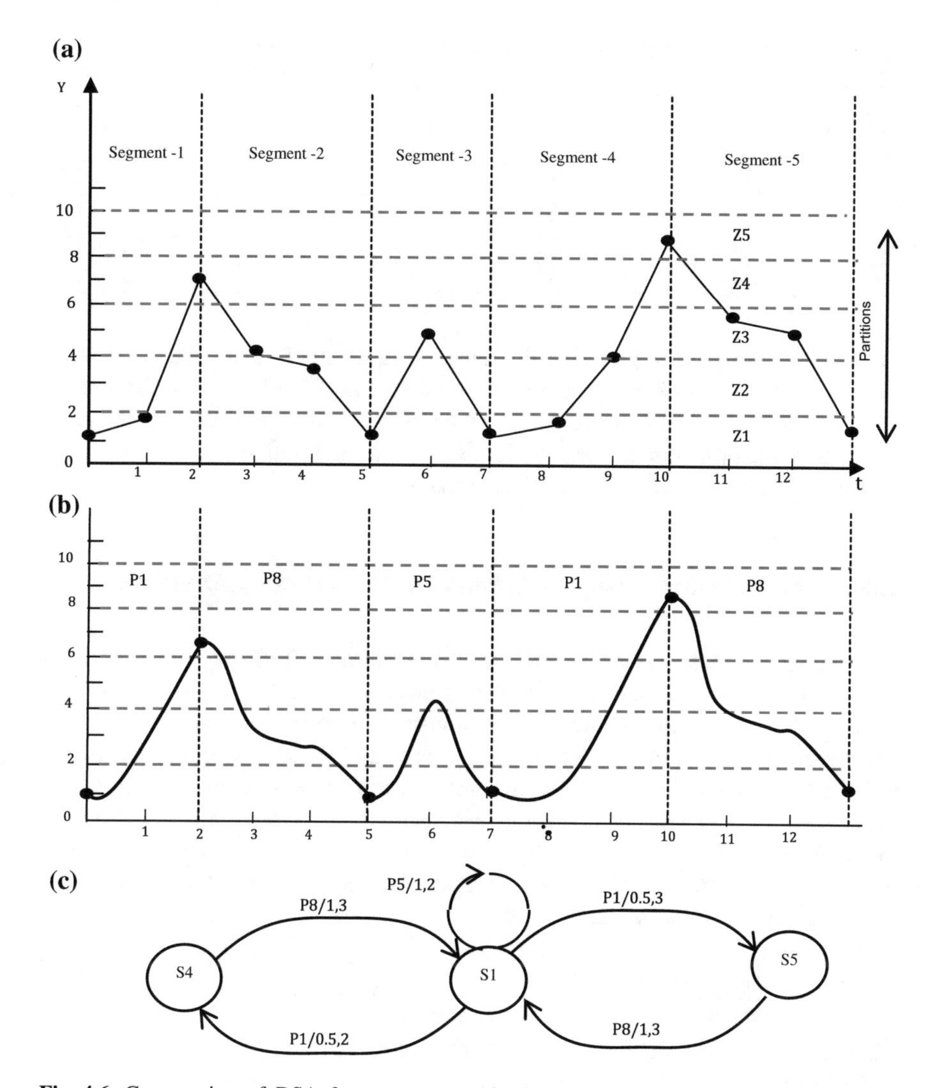


**Fig. 4.6** Construction of DSA from a segmented, clustered time-series: **a** A segmented and partitioned time-series (partitions are Z1 to Z5), **b** Segments are replaced by temporal patterns (taken from those generated for TAIEX in Fig. 4.4), **c** DSA constructed from the labeled temporal segments

where $\tau(s_i, a_k, s_j)$ represents the temporal length of the time-segment $a_k$ starting at partition $s_i$ and ending at partition $s_j$.

The approach is summarized in Pseudo Code 4.4. In Fig. 4.6, we illustrate the formation of a dynamic stochastic automaton from a segmented and clustered synthetic time-series. In Fig. 4.6a, a segmented and partitioned time-series is illustrated. In Fig. 4.6b, each temporal segment is replaced with its corresponding cluster centre. The DSA generated from the time-series is given in Fig. 4.6c.

**Table 4.1** Computation of probabilistic transition function from Fig. 4.6c

| Start state ($S_i$) | Input symbol ($p$) | End state ($S_j$) | *freq_cnt* ($S_i$, $p$, $S_j$) | $\sum$*freq_cnt* ($S_i$, $p$, $S_k$) | |
|---|---|---|---|---|---|
| S1 | P1 | S4 | 1 | 2 | (1/2) = 0.5 |
| | | S5 | 1 | | (1/2) = 0.5 |
| S1 | P5 | S1 | 1 | 1 | (1/1) = 1 |
| S4 | P8 | S1 | 1 | 1 | (1/1) = 1 |
| S5 | P8 | S1 | 1 | 1 | (1/1) = 1 |

The computations of the probabilistic transition function of the arcs in the DSA of Fig. 4.6c are given in Table 4.1. Since, in the synthesized time-series, for a given pair of starting and end states and a given pattern, there is only one temporal segment, the temporal transition function is equal to the length of the corresponding temporal segment for each arc in the automaton.

### 4.6.2 Forecasting Using the Dynamic Stochastic Automaton

Forecasting the most probable sequence of transitions (MPST) at a given starting state to reach a target state is an important problem in economic time-series prediction. Such forecasting requires three inputs: (i) a completely labelled dynamic stochastic automaton, (ii) the current state obtained from the day of prediction, and (iii) the target state. Occasionally, the users can provide a time-limit within which they expect the sequence of transitions to reach the target state. In case there exist multiple sequences of transitions to reach a given target state, then the expected duration of the most probable sequence is checked against a threshold before recommendation of the sequence to users.

Pseudo Code 4.6.1: Computation of DSA

***Input***: *A sequence* $<a_1, a_2, \ldots, a_l>$ *of l time-segments*
***Output***: *A DSA corresponding to the given temporal segments.*

*BEGIN*
*Partition the time-series into k* (=7) *horizontal partitions* $z_i$ *and construct a state* $s_i$ *for partition* $z_i$ *for* $i = 1$ *to k*;
*FOR each temporal segment* $a_k$ *do*
*BEGIN*
*IF* $s_i$ *is the starting state and* $s_j$ *is the ending state of* $a_k$ *THEN*
*Insert a directed arc from state* $s_i$ *to state* $s_j$ *in the automaton and label it with* $\delta(s_i, a, s_j)$ *and* $\delta_t(s_i, a, s_j)$ *in order following* (4.19) *and* (4.20)*; END IF;*
*END FOR;*
*END.*

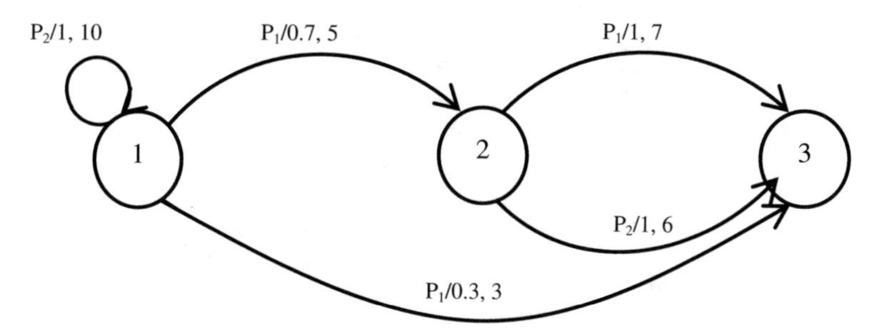


**Fig. 4.7** Example of a time-labeled dynamic stochastic

1. For a given starting state $s_1$ and a given terminal state $s_2$, determine the sequence of all possible transitions $p$ such that, $\delta'(s_1, p, s_2)$ is maximized. Let $\delta'(s_1, p, s_2) = \delta(s_1, p_1, s') \times \delta(s', p_2, s'') \times \delta(s'', p_3, s''')$. Then the most probable sequence of patterns is $<p_1, p_2, p_3>$.
2. To determine the time required for the occurrence of the MPST, we add the expected durations of each transition in $p$. Thus, the total time $T$ required, is

$$T = \sum_{\forall i} t(p_i) \tag{4.21}$$

where $p_i$ is a pattern in $p$ and $t(p_i)$ denotes the expected duration of the state transition due to pattern $p_i$, In order to realize the above approach, the following steps are carried out.
3. If $t_L$ is the allowed time-limit, we check if $T \leq t_L$. For multiple MPSTs occurring within the given time-limit, we report the sequence with the minimum expected duration.

*Example 4.8.1* Let there be a dynamic stochastic automaton as shown in Fig. 4.7.

In the automaton, we try to identify the MPST from state 1 to state 3 within a given time limit of 12 days (say). The sequences of transitions which have an expected duration within the time limit of 12 units are $<P_1, P_2>$, $<P_1, P_1>$ and $<P_1>$ respectively, as shown in Fig. 4.8. The most probable sequences of transitions from state 1 to state 3 are $<P_1, P_2>$ and $<P_1, P_1>$. However, as we have multiple sequences of most probable transitions, all occurring within the given time limit, we report the sequence having the least expected duration. Hence, we report $<P_1, P_2>$ as the most probable sequence of transitions. □

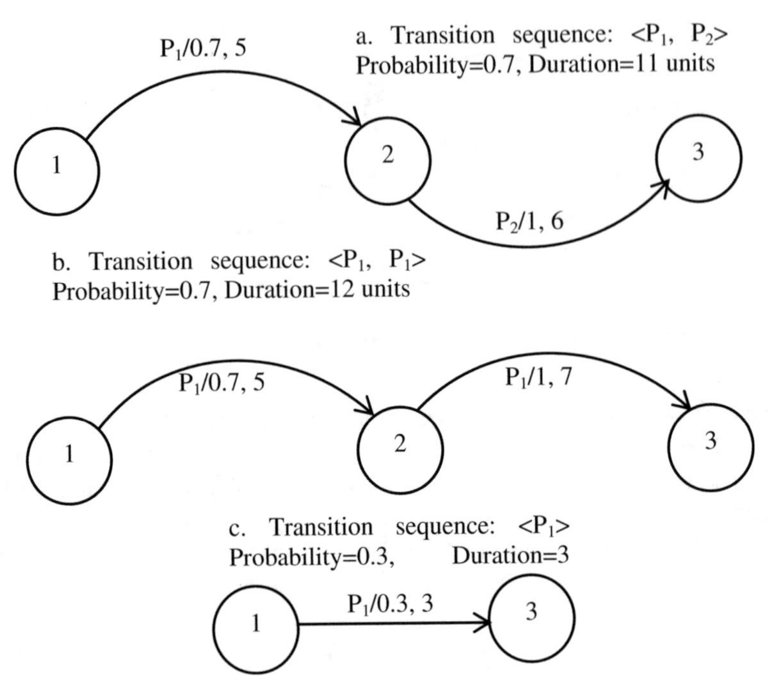


**Fig. 4.8** Possible transitions from state 1 to state 3 within given time limit

## 4.7 Computational Complexity

The first step of the SSNS segmentation algorithm, i.e., calculation of partition width requires an iteration over each pair of consecutive data points in the time-series. The partitions are stored in order, in an array. Hence, the time complexity for partitioning is $O(n)$ and the space complexity is $O(n+z)$ where $n$ is the number of data points in the time-series and $z$ is the number of partitions made.

Construction of the T-SOT involves iterating over each pair of consecutive data points in the time-series and identifying the partition to which each data point belongs. Since the array of partitions is ordered, a binary search algorithm is employed here. Hence, computing the partition to which a data point belongs takes $O(\log z)$ amount of time. This computation is done for $n$ data points and hence, the total time complexity is $O(n \log z)$. Furthermore, the windowing, window labelling and segmentation steps together, iterate over each character in the T-SOT and place a marker at each segment boundary. The window being of fixed length, the decision to place a segment boundary can be taken in constant time. As the T-SOT contains $n-1$ linguistic characters, this step has a time complexity of $O(n)$. Thus, the overall time complexity is $O(n) + O(n \log z) + O(n) \approx O(n \log z)$ and the space complexity is $O(n+z)$.

Let there be $l$ temporal segments obtained from the time-series. Each segment is approximated as a 10-dimensional point in order for it to be clustered based on the pattern that it represents. In the initialization step of the proposed multi-layered DBSCAN clustering algorithm, a distance matrix is computed from the given points. Given $l$ points, such a distance matrix can be computed in $O(l^2)$ time. In each density based stratum, the first step involves tuning the value of $\varepsilon$ corresponding to the maximum density clusters. This is done by taking the mean of the least $k$ distances from the distance matrix, where $k$ is an empirically chosen constant. Hence, the time complexity involved is $O(l^2)$.

The second step constructs the PIT, i.e., a table having $l$ entries where the $i$th entry stores the indices of points lying in the $\varepsilon$-neighbourhood of the $i$th point. This also takes $O(l^2)$ time. The third step involves removal of outliers from a density based stratum. It requires to iterate over each entry in the PIT, taking constant time to decide if a single entry (point) is an outlier. Hence, this step can be done in $O(l)$ time. The fourth step, i.e., DBSCAN clustering of the remaining points normally has a time complexity of $O(l^2)$. With the current modification of DBSCAN, searching for a core point takes $O(l)$ time. However, the PIT can be used to identify the $\varepsilon$-neighbourhood points of the core point in constant time. Thus, the total time complexity of modified-DBSCAN is also $O(l^2)$. It is important to note that the number of density based strata is significantly less compared to the number of points $l$, to be clustered and hence, can be considered as a constant. Thus the overall time-complexity for the proposed clustering algorithm is $O(l^2) + O(l^2) + O(l) + O(l^2) \approx O(l^2)$. Since, the distance matrix takes $O(l^2)$ space, the space complexity of the algorithm is also $O(l^2)$.

Having obtained the primitive patterns present in the time-series, the next task involves labelling each temporal segment to a certain cluster based on its shape, and storing the obtained knowledge in the form of a dynamic stochastic automaton. Given $z$ partitions of the time-series and $x$ primitive patterns, the automaton can be constructed in $O(z^2x)$ time. The time and space complexities of the segmentation and clustering algorithms have been summarized in Table 4.2. Also, the CPU runtime in milliseconds is mentioned, for carrying out segmentation of a time-series of 5474 data points as well as clustering the obtained segments. It is computed, by executing the segmentation and clustering algorithms on an Intel Core i5 processor machine with a CPU speed of 2.30 GHz and using MATLAB for executing the programs.

**Table 4.2** Computational and runtime complexities

| Algorithm | Time complexity | Space complexity | CPU runtime (ms) |
|---|---|---|---|
| SSNS segmentation | $O(n \log z)$ | $O(n+z)$ | 1872.0 |
| Multi-layered DBSCAN clustering | $O(l^2)$ | $O(l^2)$ | 3868.8 |

## 4.8 Prediction Experiments and Results

This section proposes an experiment to examine the success of the proposed model in forecasting, using a dynamic stochastic automaton constructed from the TAIEX economic close price time-series. The time-series is first divided into two parts, the first part to be used for knowledge and forecasting and the second part for validation. Here, the first part refers to the TAIEX close price time-series from 1st January, 1990 to 31st December 2000 (10 years) and the second part includes the period from 1st January, 2001 to 31st December 2001 (1 year). The experiment involves the following steps.

Step 1. *Automaton construction from the time-series*: The steps used for segmentation, clustering and construction of the dynamic stochastic automaton introduced earlier are undertaken.

Step 2. *Prediction of the most probable sequence of partitions*: Steps introduced in Sect. 4.6.2 are invoked for the prediction of the most probable sequence of transitions (MPST) to reach a target state from a given starting state.

Step 3. *Validation*: In the validation phase, we construct a dynamic stochastic automaton again for the entire span of 11 years from 1st January, 1990 to 31st December 2001. We forecast the most probable sequence of transitions using the automaton constructed previously in step 1 and validate our prediction using the automaton from the data of 11 years (1990–2001).

Here, we partition a time-series into seven equi-spaced partitions, namely, Extremely Low (1), Very Low (2), Low (3), Medium (4), High (5), Very High (6) and Extremely High (7). The dynamic stochastic automaton obtained from the 10 years (1990–2000) data of the TAIEX close price time-series (patterns given in Fig. 4.4) is shown in Fig. 4.9. We carry out the prediction on every day of the first 9 months (assuming a threshold time-limit of 90 days) of the year 2001. In Table 4.3, we show the probabilistic and duration accuracy of our approach by matching the probability and duration of occurrence of the MPST in both training and test phase automata. The results of certain chosen dates where the time-series changes are most prominent have been shown in Table 4.3 Prediction accuracies shown in the table and those obtained for the entire testing period are similar.

It is apparent from Table 4.3 that the average probabilistic prediction accuracy obtained by the proposed technique is very high with a margin of approximately 87.71%. In addition, the average predicted duration accuracy of the MPST on any given trading day is also very high of the order of 91.38%. Moreover, the MPST predicted in the training phase, matches with that obtained in the test phase with an accuracy of 93.52% over all the test cases in the experiment. As seen in Table 4.3, from the chosen dates, 9 out of 10 MPST predictions result in an exact match. The experimental results thus indicate that the proposed knowledge acquisition model might help in real world stock market prediction.

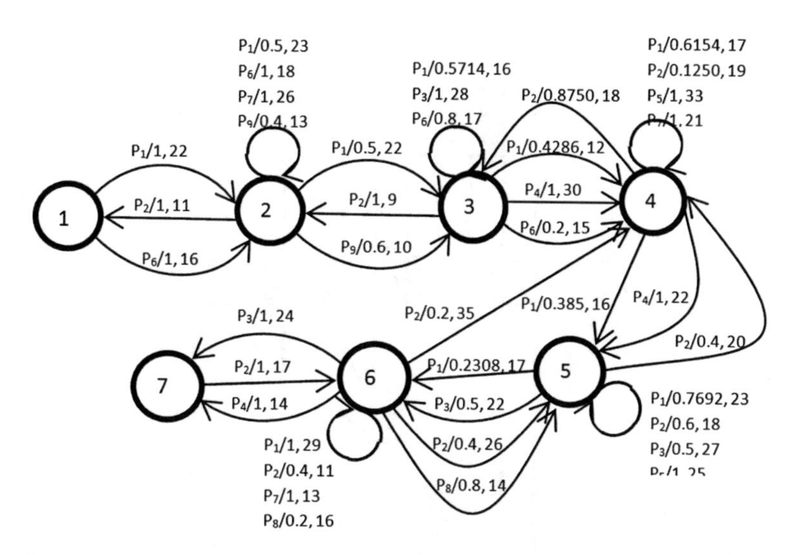


**Fig. 4.9** Dynamic Stochastic automaton obtained from TAIEX (1990–2000) time-series

## 4.9 Performance Analysis

In this section, we analyze the performance of the proposed time-series segmentation and clustering algorithms with traditional algorithms available in the literature. For segmentation, the SSNS algorithm is compared with the Top-down [26], Bottom-up [27], Sliding Window (SW) [25] and Feasible Space Window (FSW) [24] algorithms. In order to compare the performance of different segmentation algorithms, we define a metric called match-score. Two segment boundaries from different segmentation algorithms are said to match if they are placed at the same time-instant. The total number of matched segment boundaries provides the match-score of the outputs of two algorithms provided that the time-series on which they are executed is the same. As our objective for segmentation in this chapter is to identify points of significant change in the slope of the time-series, we compute the match-score of outputs obtained from different segmentation algorithms with respect to the output obtained from a hand-crafted ground-truth segmentation of the same time-series. For this analysis, we execute the above mentioned segmentation algorithms on a sample time-series of 150 data points taken from the TAIEX close-price time-series. The algorithms are coded in MATLAB and the results are summarized in Table 4.4. The outputs of the above mentioned algorithms could not be shown here due to lack of space and are given in [37] for convenience.

In order to analyse the performance of the proposed extension of DBSCAN, we compare it with the traditional k-means [18], fuzzy c-means [19] and DBSCAN [14] algorithms, using two performance indices, the *sum of squared intra-cluster distance* (SSID) and *sum of inter-cluster distances* (SICD) [17]. Let there be $m$ clusters with cluster centres $(Z_1, Z_2, \ldots, Z_m)$. Then the SSID index is computed as follows,

**Table 4.3** Prediction results obtained from TAIEX close-price time-series

| Date of prediction | Starting partition | Target partition | Prediction automaton | | | Test automaton | | | Probabilistic accuracy | Duration accuracy |
|---|---|---|---|---|---|---|---|---|---|---|
| | | | MPST | Probability ($P_P$) | Duration ($T_P$) | MPST | Probability ($P_T$) | Duration ($T_T$) | | |
| 02-02-2001 | 7 | 5 | $P_2$, $P_8$ | 0.8 | 31 | $P_2$,$P_8$ | 0.75 | 32 | 93.75 | 96.78 |
| 20-03-2001 | 6 | 5 | $P_8$ | 0.8 | 14 | $P_8$ | 0.75 | 14 | 93.75 | 100.00 |
| 01-06-2001 | 4 | 2 | $P_2$, $P_2$ | 0.8750 | 27 | $P_2$,$P_2$ | 0.9160 | 25 | 95.31 | 92.59 |
| 18-07-2001 | 2 | 3 | $P_9$ | 0.6 | 10 | $P_9$ | 0.75 | 13 | 75.00 | 70.00 |
| 14-08-2001 | 3 | 1 | $P_2$, $P_2$ | 1.0 | 20 | $P_2$,$P_2$ | 1.0 | 18 | 100.00 | 90.00 |
| *21-09-2001* | *1* | *3* | $P_6$, $P_9$ | *0.6* | *26* | $P_1$,$P_9$ | *0.5* | *33* | *N.A.* | *N.A.* |
| 01-11-2001 | 2 | 5 | $P_9$, $P_4$, $P_4$ | 0.6 | 62 | $P_9$,$P_4$, $P_4$ | 0.5120 | 58 | 85.33 | 93.55 |
| 20-11-2001 | 3 | 6 | $P_4$, $P_4$, $P_3$ | 0.5 | 74 | $P_4$,$P_4$, $P_3$ | 0.4320 | 75 | 86.40 | 98.65 |
| 10-12-2001 | 4 | 6 | $P_4$, $P_3$ | 0.5 | 52 | $P_4$,$P_3$ | 0.4320 | 56 | 86.40 | 92.31 |
| 20-12-2001 | 5 | 6 | $P_3$ | 0.5 | 30 | $P_3$ | 0.6328 | 32 | 73.44 | 88.53 |

**Table 4.4** Comparison of time-series segmentation algorithms

| Algorithm | Number of matched segment boundaries (m) | Number of segment boundaries obtained (S) | Percentage of segment boundaries matched (m/S) × 100 |
|---|---|---|---|
| Top-down [26] | 7 | 21 | 33.3 |
| Bottom-up [27] | 7 | 19 | 36.8 |
| SW [25] | 11 | 26 | 42.3 |
| FSW [24] | 9 | 18 | 50.0 |
| SSNS | 13 | 17 | 76.5 |

$$D = \sum_{i=1}^{m} \sum_{X \in S_i} \|X - Z_i\| \tag{4.22}$$

where, $S_i$ is the set of points in the $i$th cluster and $\|X - Z_i\|$ is the Euclidean distance of the point $X$ from its corresponding cluster centre $Z_i$. Clearly, a low value of the SSID index indicates a good performance of the clustering algorithm.

The SICD index is computed between cluster centres as follows,

$$D' = \sum_{i \neq j} \|Z_i - Z_j\| \tag{4.23}$$

where, $i,j \in \{1, 2, .., m\}$. The SICD index provides an estimation of inter-cluster separation. Naturally, a higher value of the SICD index indicates a well-spaced separation between individual clusters and thus a better performance of the clustering algorithm. We have executed the above mentioned clustering algorithms on the temporal segments obtained from the TAIEX close-price time-series of 1990–2000, and summarized the results in Table 4.5.

In order to compare the relative performance of the proposed extension of DBSCAN algorithm with the original DBSCAN, we execute both the algorithms (in MATLAB) for clustering the temporal segments obtained from the TAIEX close-price time-series for the period 1990–2000. The experimental results reveal that the number of clusters detected by the DBSCAN algorithm is two, representing the bullish (near-linear rise) and bearish (near-linear fall) temporal patterns as shown in Fig. 4.10. On the other hand, the proposed multi-layered extension of

**Table 4.5** Comparison of existing clustering algorithms with the proposed technique

| Algorithm / Performance index | K-means [18] | Fuzzy c-means [19] | DBSCAN [14] | Multi-layered DBSCAN |
|---|---|---|---|---|
| SSID | 15.8917 | 11.2476 | 12.5371 | 10.6215 |
| SICD | 1186.9 | 1147.4 | 1222.2 | 1293.2 |

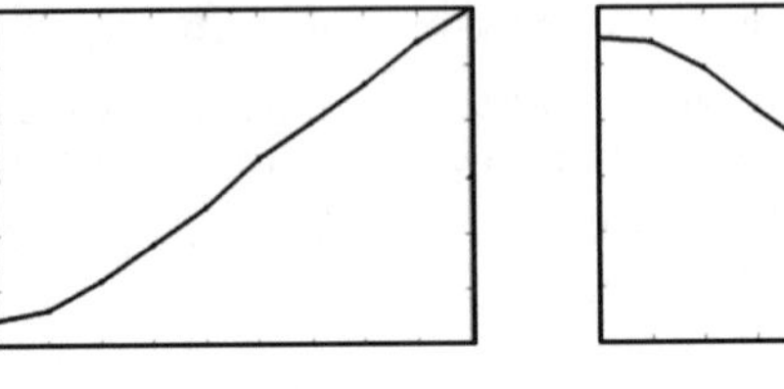
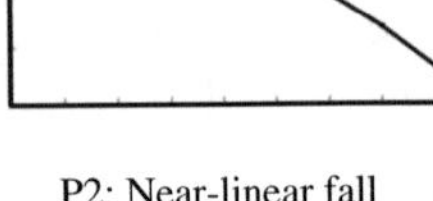


**Fig. 4.10** Temporal patterns obtained from TAIEX by executing the DBSCAN algorithm

DBSCAN is able to identify nine temporal patterns (as shown in Fig. 4.4). Also the cluster centres produced are automatically arranged in decreasing order of density, where the density of the *i*th cluster is defined is follows

$$Density = \frac{R_i}{D_i} \tag{4.24}$$

In (4.24), $R_i$ is the number of data points in the *i*th cluster and $D_i$ is the maximum intra-cluster distance between any two points in the *i*th cluster. The densities of the clusters obtained by executing both the algorithms on the said data set are summarized in Table 4.6. The results obtained from the above experiment provide conclusive proof of the advantage that the proposed extension of DBSCAN provides, whereby it is able to segregate clusters of variable densities and successfully detect temporal patterns of lower density (or lesser number of occurrences) in the data set. Since DBSCAN is not able to adapt its $\varepsilon$ parameter for clusters of variable densities, it is able to identify only the two clusters of maximum density (bullish and bearish patterns).

## 4.10 Conclusion

The merit of the chapter lies in the overall formulation of knowledge acquisition from a time-series along with the merits of individual algorithms used in the knowledge acquisition process. The segmentation algorithm is used to isolate time-blocks of similar structural characteristics, i.e., positive, negative or near-zero slopes from the rest. The clustering algorithm proposed here is an extension of the well-known DBSCAN [14] algorithm with an aim to hierarchically cluster temporal segments based on their data-density. DBSCAN is used as the base clustering algorithm for its advantage of not requiring the number of clusters as an input parameter. The merit of the extension lies in multi-level hierarchical clustering, where each level is used to cluster points of maximum data-density available in that level. The proposed extension is advantageous as it successfully captures clusters of lower data-density as well. The merit of the dynamic stochastic automaton lies in its inherent structure for knowledge representation which includes the information about both the probability of occurrence as well as the expected duration of each

**Table 4.6** Cluster densities of the patterns obtained from DBSCAN and multi-layered DBSCAN when executed on temporal segments obtained from TAIEX (1990–2000)

| DBSCAN (pattern: density of pattern) | Multi-layered DBSCAN (pattern: density of pattern) |
|---|---|
| P1: 17.8137 | P1: 31.4546 |
| | P2: 29.8410 |
| | P3: 14.5017 |
| | P4: 11.6547 |
| | P5: 7.1405 |
| P2: 16.4261 | P6: 6.0197 |
| | P7: 4.2563 |
| | P8: 3.4682 |
| | P9: 3.2773 |

transition. These parameters have been successfully applied to predict the most probable sequence of transitions (MPST) between two given states within a finite time-limit. Experiments undertaken revealed a very high prediction accuracy of over 90% (with a probabilistic accuracy of 87.71% and a duration accuracy of 91.38%) for prediction of the MPST as mentioned above. The proposed model can thus be successfully used for trading and investment in stock market.

## Appendix 4.1: Source Codes of the Programs

% MATLAB Source Code of the Main Program and Other Functions for
% Learning Structures in forecasting of Economic Time-Series
% Developed by Jishnu Mukhoti
% Under the guidance of Amit Konar and Diptendu Bhattacharya

```
%% CODE FOR SEMENTATION

%% Segmentation of Scripts

%% The script is used to load and segment a time series into time blocks.
clc; clear; close all;

%CHANGE THIS PART FOR DIFFERENT TIME SERIES
load('taiex data set.txt');
clse = taiex_data_set;
%Step 0 : Filtering the time series using a Gaussian Filter for smoothing.
clse = gaussfilt([1:length(clse)]',clse,2);
zones = horizontal_zones(clse);
```

```
%Step 1 : Divide the range of the entire time series into equi width
%partitions and find out the label (R,F,E) from each point to its next.
seq = transition_sequence(clse);

%Step 2 : Find the frequency distribution of the transitions (i.e. a region
%where the rise transition has occured more will have a higher value for
%its probability density.)
[sig1,sig2,sig3] = freq_distribution(seq);

%Step 3 : Segment the time series based on the probability distributions
%brk is a binary vector (1 = segment boundary, 0 = no segment boundary)
brk = segmentation(sig1,sig2,sig3);

%Plot the segments and the time series.
plot_segments(brk,clse);

%Preparing the training set for the clustering algorithm
num_blocks = histc(brk,1);

%Initializing the parameter set for training the clustering algorithm
param_bank = zeros(num_blocks, 10);
%start_and_end contains the starting and ending indices for each block
start_and_end = zeros(num_blocks, 2);
start_and_end(1,1) = 1;
cnt = 2;
for i = 2:length(brk)
   if(brk(i) == 1)
      start_and_end(cnt-1,2) = i-1;
      start_and_end(cnt,1) = i;
      cnt = cnt + 1;
   end
end
start_and_end(cnt-1,2) = length(brk);

for i = 1:num_blocks
   ser = clse(start_and_end(i,1):start_and_end(i,2));
   %calculating division length
   l = length(ser);
   jmp = l/10;
   inc = 0;
   for j = 1:10
      param_bank(i,j) = ser(1 + floor(inc));
      inc = inc + jmp;
   end
end
```

```
%Mean normalizing the points to help in clustering
param_bank = param_bank';
[param_bank_mn,mn,sd] = mean_normalize(param_bank);
param_bank = param_bank';
param_bank_mn = param_bank_mn';
%%%%%%%%%%%%%%%%%%%%%%%%%%
%% TransitionSequence for Close

function [ seq ] = transition_sequence( clse )
%The function produces the sequences of rise and falls which will be used
%in determining the segments of the time series.
% 1 -> rise
% 2 -> level
% 3 -> fall

zones = horizontal_zones(clse);
seq = zeros(1,1);

% for i = 2:length(clse)
%    z_prev = zone_containing_point(zones,clse(i-1));
%    z_now = zone_containing_point(zones,clse(i));
%    for j = 1:(abs(z_now-z_prev))
%       if(z_now - z_prev > 0)
%          seq = [seq, 1];
%       else
%          seq = [seq, 3];
%       end
%    end
%    if(z_now - z_prev == 0)
%       seq = [seq, 2];
%    end
% end

for i = 2:length(clse)
   z_prev = zone_containing_point(zones,clse(i-1));
   z_now = zone_containing_point(zones,clse(i));
   if(z_now - z_prev > 0)
      seq = [seq, 1];
   else
      if(z_now - z_prev < 0)
         seq = [seq, 3];
      else
         seq = [seq, 2];
      end
   end
end
```

```
seq = seq (2:end);
seq = seq';
end
%%%%%%%%%%%%%%%%%%%%%%
%% Zone Containing Points

function [ zone ] = zone_containing_point( zones, pt )
%A function to return the zone which contains a given point.

zone = 0;

for i = 1:(size(zones,1))
  if(pt >= zones(i,1) && pt < zones(i,2))
    zone = i;
    break;
  end
end

end
%%%%%%%%%%%%%%%%%%%%%%%
%% Frequency Distribution

function [ sig1, sig2, sig3 ] = freq_distribution( seq )
%The function calculates three signals based on the given sequence seq,
%which identifies the continuous frequency distribution of the occurences
%of rise fall and level transitions.

sig1 = zeros(length(seq),1);
sig2 = zeros(length(seq),1);
sig3 = zeros(length(seq),1);

for i = 3:(length(seq)-2)
  window = [seq(i-2),seq(i-1),seq(i),seq(i+1),seq(i+2)];
  w1 = histc(window,1);
  w2 = histc(window,2);
  w3 = histc(window,3);
  f1 = w1/7;
  f2 = w2/7;
  f3 = w3/7;
  sig1(i) = f1;
  sig2(i) = f2;
  sig3(i) = f3;
end
```

```
% figure;
% plot([1:length(seq)]',sig1,'g*-');
% hold on;
% plot([1:length(seq)]',sig2,'k*-');
% plot([1:length(seq)]',sig3,'r*-');

End
%%%%%%%%%%%%%%%%%%%%
%% GaussianFilter

function [ zfilt ] = gaussfilt( t,z,sigma)
%Apply a Gaussian filter to a time series
%  Inputs: t = independent variable, z = data at points t, and
%     sigma = standard deviation of Gaussian filter to be applied.
%  Outputs: zfilt = filtered data.
%
%  written by James Conder. Aug 22, 2013
%  convolution for uniformly spaced time time vector (faster) Sep 4, 2014

n = length(z); % number of data
a = 1/(sqrt(2*pi)*sigma);  % height of Gaussian
sigma2 = sigma*sigma;

% check for uniform spacing
% if so, use convolution. if not use numerical integration
uniform = false;
dt = diff(t);
dt = dt(1);
ddiff = max(abs(diff(diff(t))));
if ddiff/dt < 1.e-4
  uniform = true;
end

if uniform
  filter = dt*a*exp(-0.5*((t - mean(t)).^2)/(sigma2));
  i = filter < dt*a*1.e-6;
  filter(i) = [];
  zfilt = conv(z,filter,'same');
else
  %%% get distances between points for proper weighting
  w = 0*t;
  w(2:end-1) = 0.5*(t(3:end)-t(1:end-2));
  w(1) = t(2)-t(1);
  w(end) = t(end)-t(end-1);
```

```
%%% check if sigma smaller than data spacing
iw = find(w > 2*sigma, 1);
if ~isempty(iw)
   disp('WARNING: sigma smaller than half node spacing')
   disp('May lead to unstable result')
   iw = w > 2.5*sigma;
   w(iw) = 2.5*sigma;
   % this correction leaves some residual for spacing between 2-3sigma.
   % otherwise ok.
   % In general, using a Gaussian filter with sigma less than spacing is
   % a bad idea anyway...
end

%%% loop over points
zfilt = 0*z;  % initalize output vector
for i = 1:n
   filter = a*exp(-0.5*((t - t(i)).^2)/(sigma2));
   zfilt(i) = sum(w.*z.*filter);
end

%%% clean-up edges - mirror data for correction
ss = 2.4*sigma;  % distance from edge that needs correcting

% left edge
tedge = min(t);
iedge = find(t < tedge + ss);
nedge = length(iedge);
for i = 1:nedge;
   dist = t(iedge(i)) - tedge;
   include = find( t > t(iedge(i)) + dist);
   filter = a*exp(-0.5*((t(include) - t(iedge(i))).^2)/(sigma2));
   zfilt(iedge(i)) = zfilt(iedge(i)) + sum(w(include).*filter.*z(in-
clude));
end

% right edge
tedge = max(t);
iedge = find(t > tedge - ss);
nedge = length(iedge);
for i = 1:nedge;
   dist = tedge - t(iedge(i));
   include = find( t < t(iedge(i)) - dist);
   filter = a*exp(-0.5*((t(include) - t(iedge(i))).^2)/(sigma2));
   zfilt(iedge(i)) = zfilt(iedge(i)) + sum(w(include).*filter.*z(in-
clude));
```

```
   end
end      % uniform vs non-uniform

end

%%%%%%%%%%%%%%%%%%%%
%% Horizontal Zones

function [ zones ] = horizontal_zones( clse )
%Given the time series clse, the function calculates the width of the zones
%for appropriate segmentation of the time series.

%variable to store the average difference between points
diff = 0;

for i = 2:length(clse)
   diff = diff + abs(clse(i) - clse(i-1));
end

w = diff/length(clse);

num_zones = floor((max(clse) - min(clse))/w);
zones = zeros(num_zones,2);

cnt = min(clse);
for i = 1:num_zones
   zones(i,1) = cnt;
   zones(i,2) = cnt + w;
   cnt = cnt + w;
end
end
%%%%%%%%%%%%%%%%%%%%
%% Mean Normalization

function [ X_mn,mn,sd ] = mean_normalize( X )
%A function to mean normalize the data set given in X.

%col here represents the number of columns in X which is equal to the number
%of features or the number of dimensions of a point.
%row represents the number of rows or the number of col dimensional points.
col = size(X,2);

mn = mean(X);
sd = std(X);
```

```
X_mn = zeros(size(X));

for i = 1:col
  X_mn(:,i) = (X(:,i)-mn(i))/sd(i);
end

end
%%%%%%%%%%%%%%%%%%%%%
%% Plot Segments

function [ ] = plot_segments( brk, clse )
%The function plots the segments of the time series clse as given by the
%segmentation function

figure;
len = length(clse);

plot([1:len]',clse,'r-');
hold on;

yL = get(gca,'Ylim');

for i = 1:len
  if(brk(i) == 1)
    line([i,i],yL,'Color','k');
  end
end

end
%%%%%%%%%%%%%%%%%%%%%%%%%
%% Segmentation

function [ brk ] = segmentation( sig1,sig2,sig3 )
%The function which returns the indices of the time series at which it has
%to be segmented.

%determining the length of the three signals
len = length(sig1);
%the brk vector contains the indices at which the time series is to be
%segmented.
brk = zeros(len+1,1);

brk(1) = 1;
```

```
sg = [sig1(1), sig2(1), sig3(1)];
[mx, st] = max(sg);

for i = 2:len
  sg = [sig1(i), sig2(i), sig3(i)];
  [mx,ch] = max(sg);
  if(ch == st)
    continue;
  else
    if(ch ~= 1 && st == 1 && sig1(i) == mx)
      continue;
    end
    if(ch ~= 2 && st == 2 && sig2(i) == mx)
      continue;
    end
    if(ch ~= 3 && st == 3 && sig3(i) == mx)
      continue;
    end
    brk(i) = 1;
    st = ch;
  end
end
%removing the segmentations which are within the 7 day period
k = 0;
for i = 1:(len+1)
  if(brk(i) == 1 && k == 0)
    k = 7;
  else if(brk(i) == 1 && k ~= 0)
      brk(i) = 0;
    end
  end
  if(k > 0)
    k = k - 1;
  end
end
end
%%%%%%%%%%%%%%%%%%%%

%% CODE FOR CLUSTERING

%% Main Program
%%Script to find the cluster centroids in the given time-segments using a
%%non-parametric DBSCAN clustering approach.
clc; close all;
```

```
load('param_bank.mat');
dt = param_bank_mn;
cent = repeated_cluster(dt);

n = size(cent,1);
for i=1:n
  figure;
  plot([1:10],cent(i,:),'k*-');
end

idx = assign_cluster(param_bank_mn,cent);
%%%%%%%%%%%%%%%%%%

%% Assign Cluster

function [ idx ] = assign_cluster( blocks, centroids )
%The function takes in a set of time blocks each block being a ten
%dimensional point.

num_blocks = size(blocks,1);
num_centroids = size(centroids,1);
idx = zeros(num_blocks,1);

for i = 1:num_blocks
  bl = blocks(i,:);
  c = centroids(1,:);
  min_dist = euclid_dist(bl,c);
  idx(i) = 1;
  for j = 2:num_centroids
    c = centroids(j,:);
    e = euclid_dist(bl,c);
    if e < min_dist
      min_dist = e;
      idx(i) = j;
    end
  end

end

end
%%%%%%%%%%%%%%%%%%%%%
%% Clustering

function [ processed_points ] = cluster( func, ad_list )
%A function to perform the non-parametric clustering using a depth-first
```

```
%search or seed filling approach.

%initialising the cluster count to 1
cluster_cnt = 1;

%getting the number of points to create the hash table to store cluster
%values of processed points.
m = size(func,1);
processed_points = zeros(m,1);

%setting the threshold max for the func vector which, when encountered
%stops the processing.
[mx_idx,mx] = find_max(func);

%Creating stack data structure
stack = zeros(m,1);%an array of point indices
top = 1;%top of stack
%Stack data structure created

%loop ending criterion
while(mx_idx ~= -1 && mx > 3)
   %in one iteration of this while loop we are supposed to create and form
   %one cluster from the data
   %Pushing the index onto the stack
   stack(top) = mx_idx;
   top = top + 1;
   %Continue while stack is not empty
   while(top ~= 1)
     %Pop a point from the stack
     pt_idx = stack(top-1);
     top = top - 1;
     %Continue processing on this point if this point has not been seen
     %before
     if func(pt_idx) ~= -1
        %Get the surrounding points for the current popped point
        surr = ad_list(pt_idx,:);
        counter = 1;
        %For each point in the surroinding points, push it in the stack
        %if it has not been processed.
        while 1>=0
           if surr(counter) == 0
              break;
           end
           if processed_points(surr(counter)) == 0
              stack(top) = surr(counter);
```

```
            top = top + 1;
          end
          counter = counter + 1;
        end
        %Process the point
        processed_points(pt_idx) = cluster_cnt;
        %Removing the point which has just been processed from the
        %func
        func(pt_idx) = -1;%Logical deletion
      end
    end
    %Incrementing the cluster count by 1
    cluster_cnt = cluster_cnt + 1;
    %Finding the max point and its index
    [mx_idx,mx] = find_max(func);
end

end
%%%%%%%%%%%%%%%%%%%%%%%%
%% Distribution Function

function [ nearby_points, ad_list ] = distribution_function( X )
%A function to find the number of nearby points given a particular point in
%the points matrix X.
%Each row in X corresponds to an n dimensional point where n is the number
%of columns.

num_points = size(X,1);
nearby_points = zeros(num_points, 1);
ad_list = zeros(num_points, num_points-1);
ad_list_index = ones(num_points,1);

r1 = find_rad_of_influence(X);
%fprintf('The radius chosen is : %d\n',r1);
% r = r/3;

for i = 1:num_points
    p1 = X(i,:);
    for j = 1:num_points
        if(j ~= i)
            p2 = X(j,:);
            if(euclid_dist(p1,p2) < r1)
                nearby_points(i) = nearby_points(i) + 1;
                ad_list(i,ad_list_index(i)) = j;
                ad_list_index(i) = ad_list_index(i) + 1;
```

```
            end
        end
    end
end

avr = mean(nearby_points);

%Removing all those points whose surroindings are relatively sparse
for i = 1:num_points
    if(nearby_points(i) < avr)
        nearby_points(i) = -1;
    end
end

end
%%%%%%%%%%%%%%%%%%%%%%
%% Euclidean distance

function [ d ] = euclid_dist( p1, p2 )
%A function to calculate the euclidean distance between two
%multidimensional points p1 and p2.
%p1 and p2 can be a 1*n matrix where n is the dimension of the point.

n = size(p1,2);
d = 0;
for i = 1:n
    d = d + (p1(i) - p2(i))^2;
end
d = sqrt(d);

end
%%%%%%%%%%%%%%%%%%%%%%
%% Find Centroids

function [ c ] = find_centroids( processed_points,points )
%A function to detect the centroids from the clustered points

num_clusters = max(processed_points);
dim = size(points,2);

c = zeros(num_clusters,dim);

for i = 1:num_clusters
    cnt = 0;
    for j = 1:length(processed_points)
```

```
      if(processed_points(j) == i)
        c(i,:) = c(i,:) + points(j,:);
        cnt = cnt + 1;
      end
    end
    c(i,:) = c(i,:)/cnt;
end
end
%%%%%%%%%%%%%%%%%%%%%%
%% Find Maximum

function [ idx,max ] = find_max( func )
%A function to return the index at which the maximum value of func is
%encountered.

max = 0;
max_idx = 1;
flag = 0;

for i = 1:length(func)
   if(func(i) ~= -1)
     flag = 1;
     if(func(i) > max)
       max = func(i);
       max_idx = i;
     end
   end
end

if (flag == 0)
   idx = -1;
else
   idx = max_idx;
end

end
%%%%%%%%%%%%%%%%%%%%%%%%
%% Find Radius of influence

function [ r1,dist_vec ] = find_rad_of_influence( points )
%A function to determine the radius of influence given the set of mean
%normalized points.

num_points = size(points,1);
cnt = 1;
dist_vec = zeros(num_points^2 - num_points,1);
```

```
for i = 1:num_points
   for j = 1:num_points
      if i ~= j
         dist_vec(cnt) = euclid_dist(points(i,:),points(j,:));
         cnt = cnt + 1;
      end
   end
end

dist_vec = sort(dist_vec);
len = length(dist_vec)/10;
p = dist_vec(1:len);
r1 = mean(p);

% sort(dist_vec);
% mean_rad = 0;
% cnt = 0;
% for i = 1:length(dist_vec)
%    if(dist_vec(i)>0 && dist_vec(i)<1)
%       cnt = cnt + 1;
%       mean_rad = mean_rad + dist_vec(i);
%       if cnt == 10
%          break;
%       end
%    end
% end
%
% mean_rad = mean_rad/cnt;
% r = mean_rad;
%r = mean_rad/1.5;
%fprintf('Value of cnt : %d\n',cnt);

end
%%%%%%%%%%%%%%%%%%%%%%%%%%%
%%Mukhoti Clustering

function [ processed_points,c,outliers ] = mukhoti_clustering( points )
%The brand new clustering algorithm invented by yours truly.

pt = points;
[func, ad_list] = distribution_function(pt);

processed_points = cluster(func,ad_list);
c = find_centroids(processed_points,points);
```

```
outliers = zeros(histc(func,-1),size(points,2));
cnt = 1;
for i = 1:size(points,1);
  if(func(i) == -1)
    outliers(cnt,:) = points(i,:);
    cnt = cnt + 1;
  end
end

%proc = assign_unclustered_points(c,processed_points,points);
%plot_clusters(points,processed_points);

%the final step is to find the cluster centroids and assign all the
%unassigned clusters to their centroids.
% num_clusters = max(processed_points);


end
%%%%%%%%%%%%%%%%%%%%%%%%%
%% Purge Centroids

function [ mod_centroids ] = purge_centroids( centroids, points )
%This function processes the centroids and removes the centroids which are
%too close to each other.

r = find_rad_of_influence(points);
null_ct = ones(1,size(centroids,2));
null_ct = -null_ct;

for i = 1:size(centroids,1)
  c1 = centroids(i,:);
  if sum(c1 - null_ct) ~= 0
    for j = 1:size(centroids,1)
      if j ~= i
        c2 = centroids(j,:);
        if euclid_dist(c1,c2) < (r*2)
          centroids(j,:) = null_ct;
        end
      end
    end
  end
end

mod_centroids = zeros(1,size(centroids,2));
```

```
for i = 1:size(centroids,1)
   c1 = centroids(i,:);
   if sum(c1 - null_ct) ~= 0
      mod_centroids = [mod_centroids;c1];
   end
end

mod_centroids = mod_centroids(2:end,:);

end
%%%%%%%%%%%%%%%%%%%%%
%% Repeated Clusters

function [ cent ] = repeated_cluster( points )
%Implementation of multi-layered clustering
orig_numpt = size(points,1);

[idx,c,out] = mukhoti_clustering(points);
cent = c;

num_out = size(out,1);

while num_out > (orig_numpt/10)
   [idx,c,out] = mukhoti_clustering(out);
   cent = [cent;c];
   num_out = size(out,1);
end

%Removing the redundant centroids
cent = purge_centroids(cent,points);
end
%%%%%%%%%%%%%%%%%%

%% CODE FOR AUTOMATON

%% Main Program

%%Main script to construct the dynamic stochastic automaton from the given
%%input parameters.

clear; close all; clc;

load('clse.mat');
load('idx.mat');
load('start_and_end.mat');
```

```
auto = automaton(clse,start_and_end,idx);
%%%%%%%%%%%%%%%%%%%%%%

%% Automaton

function [ automaton, test_automaton ] = automaton( clse, start_and_end,
idx )
%Generates training and test phase dynamic stochastic automata
zones = final_partitions(clse);
zones
%FOR TAIEX
test = start_and_end(105:118,:);
start_and_end = start_and_end(1:104,:);
num_partitions = size(zones,1);
fprintf('Number of partitions : %d\n',num_partitions);
num_segments = size(start_and_end,1);
fprintf('Number of segments : %d\n',num_segments);
num_patterns = max(idx);
fprintf('Number of patterns : %d\n',num_patterns);
automaton = zeros(num_partitions, num_partitions, num_patterns);

for i = 1:num_segments
  st = start_and_end(i,1);
  en = start_and_end(i,2);
  st1 = clse(st);
  en1 = clse(en);
  pst = zone_containing_point(zones,st1);
  pen = zone_containing_point(zones,en1);
  if(pst == 0)
    pst = pst + 1;
  end
  if(pen == 0)
    pen = pen + 1;
  end
%   pst = pst + 1;
%   pen = pen + 1;
  pattern = idx(i);
  automaton(pst,pen,pattern) = automaton(pst,pen,pattern) + 1;
end

for i = 1:num_partitions %for all start states
  for j = 1:num_patterns %for all input patterns
    s = sum(automaton(i,:,j));
    if (s ~= 0)
      automaton(i,:,j) = automaton(i,:,j) ./ s;
```

```
      end
    end
  end

  num_test_seg = size(test,1);
  fprintf('Number of test segments : %d\n',num_test_seg);
  test_automaton = zeros(num_partitions,num_partitions,num_patterns);

  for i = 1:num_test_seg
    st = test(i,1);
    en = test(i,2);
    st1 = clse(st);
    en1 = clse(en);
    pst = zone_containing_point(zones,st1);
    pen = zone_containing_point(zones,en1);
%   pst = pst + 1;
%   pen = pen + 1;
%
    if(pst == 0)
      pst = pst + 1;
    end
    if(pen == 0)
      pen = pen + 1;
    end
    pattern = idx(i+212);
    test_automaton(pst,pen,pattern) = test_automaton(pst,pen,pat-
tern) + 1;
end
for i = 1:num_partitions
  for j = 1:num_patterns
    s = sum(test_automaton(i,:,j));
    if (s ~= 0)
      test_automaton(i,:,j) = test_automaton(i,:,j) ./ s;
    end
  end
end
end
%%%%%%%%%%%%%%%%%%
```

```
%% Final Partitions

function [ zones ] = final_partitions( clse )
%Dividing the time-series into 7 partitions

M = max(clse);
m = min(clse);

w = (M-m)/7;

zones = zeros(7,2);

zones(1,1) = m;
zones(1,2) = m + w;
zones(2,1) = m + w;
zones(2,2) = m + (2*w);
zones(3,1) = m + (2*w);
zones(3,2) = m + (3*w);
zones(4,1) = m + (3*w);
zones(4,2) = m + (4*w);
zones(5,1) = m + (4*w);
zones(5,2) = m + (5*w);
zones(6,1) = m + (5*w);
zones(6,2) = m + (6*w);
zones(7,1) = m + (6*w);
zones(7,2) = M;

end
%%%%%%%%%%%%%%%%%%%%

%% Zone Containing Points

function [ zone ] = zone_containing_point( zones, pt )
%A function to return the zone which contains a given point.

zone = 0;
for i = 1:(size(zones,1))
   if(pt >= zones(i,1) && pt < zones(i,2))
      zone = i;
      break;
   end
end
end
%%%%%%%%%%%%%%%%%%
```

**Steps to run the above program**

1. Segmentation
   Input: TAIEX close price file
   Output: 239 × 10 matrix containing temporal segments (param_bank_mn)
   Script to run: segmentation_script
2. Clustering
   Input: Load the param_bank_mn.mat file.
   Output: Cluster centroids (cent) and cluster labels of each temporal segment (idx)
   Script to run: main_prog
3. Dynamic Stochastic Automaton
   Input: idx.mat, clse.mat, start_and_end.mat
   Output: dynamic stochastic automaton in the form of a 3d matrix (auto)
   Script to run: run_automaton

**Exercises**

1. Given the time-series, determine the segments of the series graphically using the proposed segmentation algorithm.

Justify the results of the above segmentation steep rise, fall and their frequency count.

[**Hints**: The segments are graphically obtained based on positivity, negativity and zero values in slope of the series. Small changes in sign of slope are ignored in Fig. 4.11b]

2. Given 3 sets of data with different density levels.

(a) By DBSCAN algorithm, cluster the data points into two clusters (Fig. 4.12).
[**Hints**: Only high density × points will be clustered.]
(b) What would be the response of hierarchical DBSCAN algorithm to the given 20 data.
[**Hints**: All the three clusters would be separated in multiple steps.]

3. In stochastic automaton, given below, find the sequence with the highest probability of occurrences (Fig. 4.13).

[**Hints**: The sequences beginning at $P_1$ and terminating at $P_4$ are (Fig. 4.14):
The probability of sequence bc = $1 \times 1 = 1.0$ is the highest.]

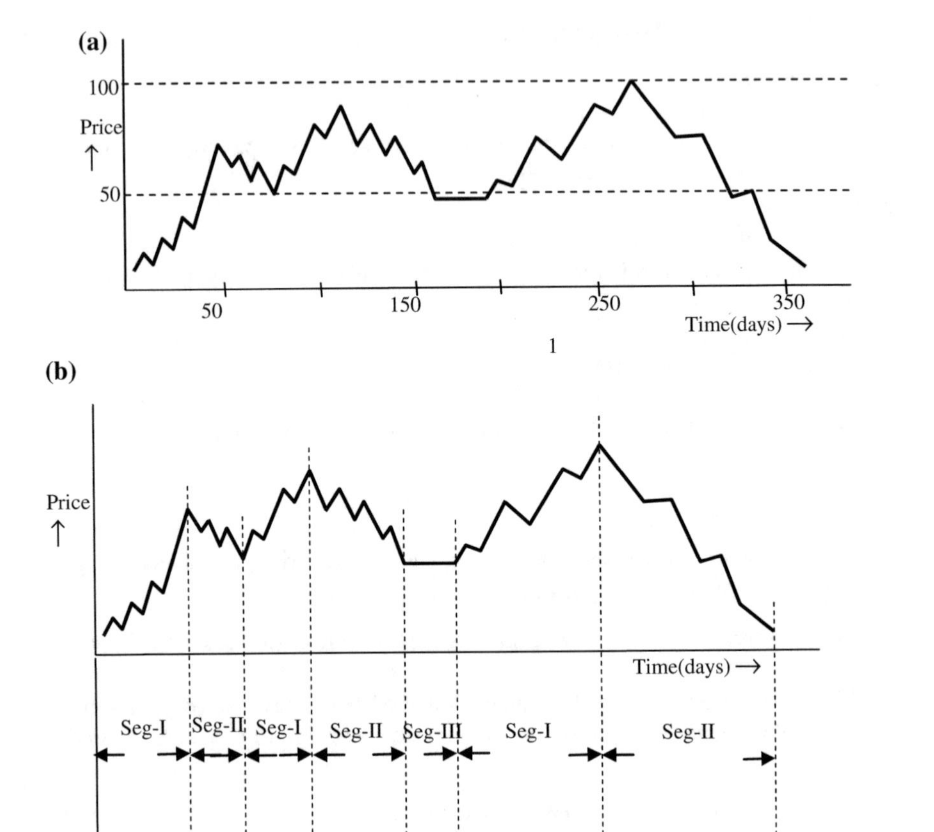


**Fig. 4.11** **a**, **b** Figure for Q1

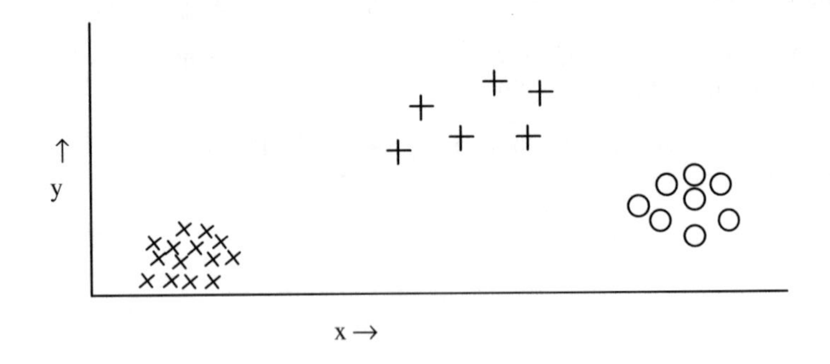


**Fig. 4.12** Figure for Q2

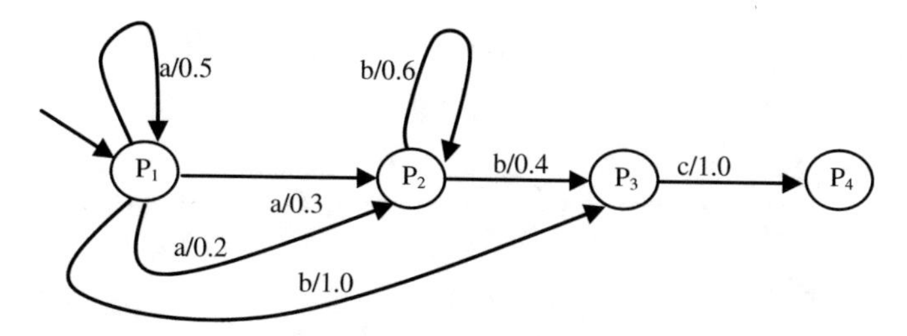


**Fig. 4.13** Figure for Q3

4. In the automaton provided in Fig. 4.15, suppose we have 4 patterns, to be introduced in the automaton for state transition. If today's price falls in partition $P_4$, what would be the next appropriate time required for the state transition? Also what does prediction infer?

[**Hints**: Next state $P_5$, appropriate time required is 5 days. The prediction infers growth following cluster I.]

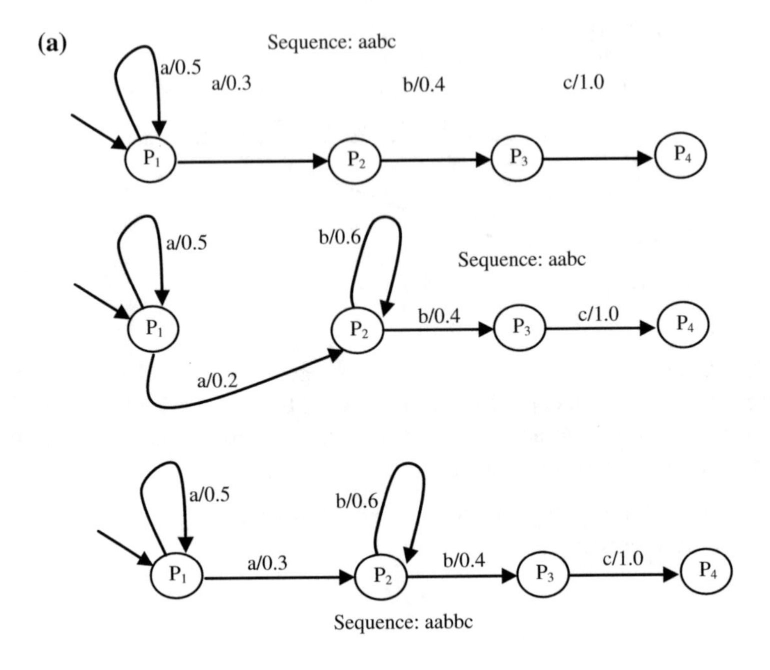


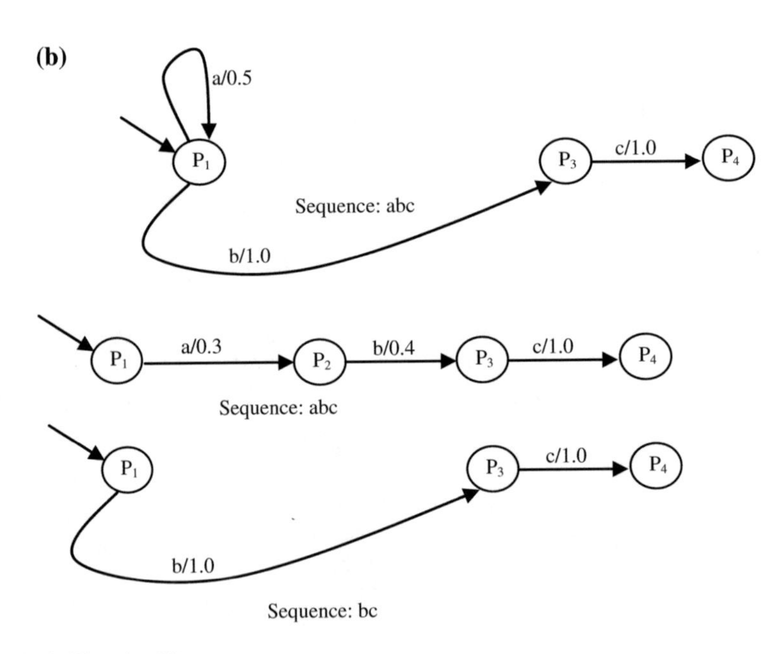


**Fig. 4.14** Hints for Q3

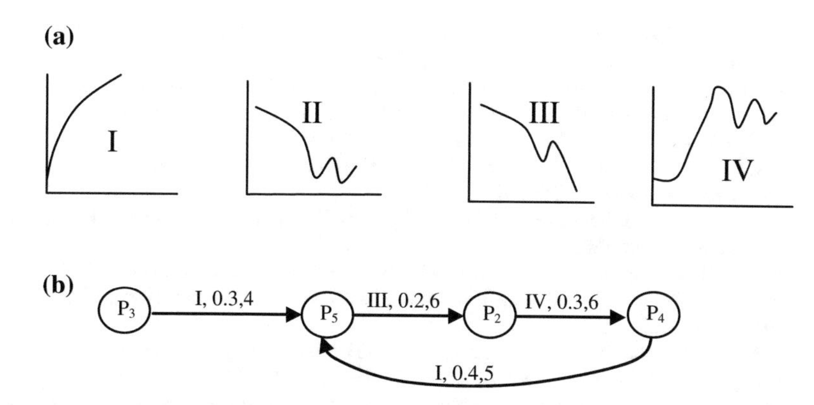


**Fig. 4.15** **a** Clustered pattern used for prediction. **b** The automata representing transitions from one partition to the other with clustered patterns as input for state transition, the transition function includes input cluster, probability of transition and appropriate number of days required for transition in order

# References


1. Barnea, O., Solow, A. R., & Stone, L. (2006). On fitting a model to a population time series with missing values. *Israel Journal of Ecology and Evolution, 52,* 1–10.
2. Chen, S. M., & Hwang, J. R. (2000). Temperature prediction using fuzzy time series. *IEEE Transactions on Systems, Man, and Cybernetics, Part-B, 30*(2), 263–275.
3. Wu, C. L., & Chau, K. W. (2013). Prediction of rainfall time series using modular soft computing methods. *Elsevier, Engineering Applications of Artificial Intelligence, 26,* 997–1007.
4. Jalil, A., & Idrees, M. (2013). Modeling the impact of education on the economic growth: Evidence from aggregated and disaggregated time series data of Pakistan. *Elsevier Economic Modelling, 31,* 383–388.
5. Song, Q., & Chissom, B. S. (1993). Fuzzy time series and its model. *Elsevier, Fuzzy Sets Systems, 54*(3), 269–277.
6. Song, Q., & Chissom, B. S. (1993). Forecasting enrollments with fuzzy time series—Part I. *Elsevier, Fuzzy Sets Systems, 54*(1), 1–9.
7. Mirikitani, D. T., & Nikolaev, N. (2010). Recursive Bayesian recurrent neural networks for time-series modeling. *IEEE Transaction on Neural Networks, 2*(2), 262–274.
8. Connor, J. T., Martin, R. D., & Atlas, L. E. (1994). Recurrent neural networks and robust time series prediction. *IEEE Transaction on Neural Networks, 5*(2), 240–254.
9. De Gooijer, J. G., & Hyndman, R. J. (2006). 25 years of time series forecasting. *Elsevier, International Journal of Forecasting, 22,* 443–473.
10. Lee, C. H. L., Liu, A., & Chen, W. S. (2006). Pattern discovery of fuzzy time series for financial prediction. *IEEE Transactions on Knowledge and Data Engineering, 18*(5), 613–625.
11. Bhattacharya, D., Konar, A., & Das, P. (2016). Secondary factor induced stock index time-series prediction using self-adaptive interval type-2 fuzzy sets. *Elsevier, Neurocomputing, 171,* 551–568.
12. Last, M., Klein, Y., & Kandel, A. (2001). Knowledge discovery in time series databases. *IEEE Transactions on Systems, Man, and Cybernetics, Part B, 31*(1), 160–169.
13. Dougherty, E. R., & Giardina, C. R. (1988). *Mathematical methods for artificial intelligence and autonomous systems*. USA: Prentice-Hall, The University of Michigan.
14. Ester, M., Kriegel, H. P., Sander, J., & Xu, X. (1996). A density-based algorithm for discovering clusters in large spatial databases with noise. In *Proceedings of KDD-96* (pp. 226–231).

15. Konar, A. (1999). *Artificial intelligence and soft computing: Behavioural and cognitive modeling of the human brain*. Boca Raton, Florida: CRC Press.
16. TAIEX [Online]. Available http://www.twse.com.tw/en/products/indices/tsec/taiex.php
17. Bezdek, J. C., & Pal, N. R. (1998). Some new indexes of cluster validity. *IEEE Transactions on Systems, Man, and Cybernetics, Part B, 28*(3), 301–315.
18. Forgy, E. W. (1965). Cluster analysis of multivariate data: Efficiency versus interpretability of classification. *Biometrics, 21*(3), 768–769.
19. Bezdek, J. C. (1987). *Pattern recognition with fuzzy objective function algorithms*. New York and London: Plenum Press.
20. Edwards, R. D., Magee, J., & Bassetti, W. H. C. (2001). The dow theory. In *Technical analysis of stock trends* (8th ed., 2000, pp. 21–22). N.W. Corporate Blvd., Boca Raton, Florida: CRC Press LLC.
21. Keogh, E., Chu, S., Hart, D., & Pazzani, M. (2003). Segmenting time series: A survey and novel approach. *Data mining in time series databases* (2nd ed.). World Scientific.
22. Keogh, E., & Pazzani, M. (1999). Scaling up dynamic time warping to massive dataset. In *Proceedings of third European conference principles of data mining and knowledge discovery* (PKDD '99) (pp. 1–11).
23. Kanth, K. V. R., Agrawal, D., & Singh, A. K. (1998). Dimensionality reduction for similarity searching in dynamic databases. In *Proceedings of ACM SIGMOD '98* (pp. 166–176).
24. Liu, X., Lin, Z., & Wang, H. (2008). Novel online methods for time series segmentation. *IEEE Transactions of Knowledge and Data Engineering, 20*(12), 1616–1626.
25. Appel, U., & Brandt, A. V. (1983). Adaptive sequential segmentation of piecewise stationary time series. *Information Science, 29*(1), 27–56.
26. Shatkay, H., & Zdonik, S. B. (1995). *Approximate queries and representations for large data sequences*. Technical report CS-95-03, Department of Computer Science, Brown University.
27. Keogh, E., & Smyth, P. (1997). A probabilistic approach to fast pattern matching in time series databases. In *Proceedings of ACM SIGKDD'97* (pp. 20–24).
28. Bryant, G. F., & Duncan, S. R. (1994). A solution to the segmentation problem based on dynamic programming. In *Proceedings of third IEEE conference of control applications (CCA '94)* (pp. 1391–1396).
29. Abonyi, J., Feil, B., Nemeth, S., & Arva, P. (2005). Modified Gath-Geva clustering for fuzzy segmentation of multivariate time-series. *Elsevier, Fuzzy Sets and Systems, 149*(1), 39–56.
30. Fuchs, E., Gruber, T., Nitschke, J., & Sick, B. (2010). Online segmentation of time series based on polynomial least-squares approximations. *IEEE Transactions on Pattern Analysis, Machine Intelligence, 32*(12), 2232–2245.
31. Chung, F. L., Fu, T. C., Ng, V., & Luk, R. W. P. (2004). An evolutionary approach to pattern-based time series segmentation. *IEEE Transactions of Evolutionary Computation, 8*(5), 471–489.
32. Xu, R., & Wunsch, D., II. (2005). Survey of clustering algorithms. *IEEE Transactions on Neural Networks, 16*(3), 645–678.
33. Mitchell, T. M. (1997). *Machine learning*. New York, USA: McGraw-Hill.
34. Liao, T. W. (2005). Clustering of time series data—a survey. *Elsevier, Pattern Recognition, 38*(11), 1857–1874.
35. Xiong, Z., Chen, R., Zhang, Y., & Zhang, X. (2012). Multi-density DBSCAN algorithm based on density levels partitioning. *Journal of Information & Computational Science, 9*(10), 2739–2749.
36. Lundberg, J. (2007). Lifting The Crown-Citation Z-score. *Elsevier, Journal of Informatics, 1,* 145–154.
37. http://computationalintelligence.net/tnnls/segmentation_performance.pdf

# Chapter 5
# Grouping of First-Order Transition Rules for Time-Series Prediction by Fuzzy-Induced Neural Regression

**Abstract** In this chapter, we present a novel grouping scheme of first-order transition rules obtained from a partitioned time-series for fuzzy-induced neural regression. The transition rules here represent precedence relationships between a pair of partitions containing consecutive data points in the time-series. In this regard, we propose two neural network ensemble models. The first neural model represents a set of transition rules, each with a distinct partition in the antecedent. During the prediction phase, a number of neural networks containing the partition corresponding to the current time-series data point in the antecedent are triggered to produce outputs following the pre-trained rules. Pruning of rules that do not contain the partition corresponding to the current data point in the antecedent is performed by a pre-selector Radial Basis Function neural network. In the first model, the partitions present in transition rules are described by their respective mid-point values during neural training. This might induce approximation error due to representation of a complete band of data points by their respective partition mid-points. In the second model, we overcome this problem by representing the antecedent of a transition rule as a set of membership values of a data point in a number of fuzzy sets representing the partitions. The second model does not require selection of neural networks by pre-selector RBF neurons. Experiments undertaken on the Sunspot time-series as well as on the TAIEX economic close-price time-series reveal a high prediction accuracy outperforming competitive models, thus indicating the applicability of the proposed methods to real life time-series forecasting.

## 5.1 Introduction

A time-series is a discrete sequence of values obtained from a time-varying function. Generally, a time-series is one of the preferred representations for analytical study of various natural phenomena like atmospheric temperature [1], rainfall [2], seismic activity [3], population growth [4], economic growth [5] and the like. The main objective behind time-series analysis is to make accurate predictions on



A. Konar and D. Bhattacharya, *Time-Series Prediction and Applications*,
Intelligent Systems Reference Library 127,
DOI 10.1007/978-3-319-54597-4_5

possible future values of the series. The task of forecasting is difficult because the actual factors affecting a time-series are manifold and often unknown. Hence, the conventional procedure adopted by researchers is to develop a model which uses values of the time-series at time instants $(t-k),(t-k+1),\ldots,t-2,t-1,t$ for some positive integer k to make a prediction on the value of the series at time instant $t+1$.

A large number of statistical methods for time-series modeling and prediction exist in the literature [6–9]. Among these models the Autoregressive Integrated Moving Average (ARIMA) [7] needs special mention. The ARIMA model combines both the autoregressive model [8] and the moving average model [9] to obtain better forecasts. It is based on three parameters, namely, $p$, the order of auto regression, $d$, the degree of differencing and $q$, the order of the moving average model. It defines the time-series value at a particular time instant as the sum of a linear combination of earlier time-series values along with a linear combination of earlier error terms (difference between predicted values and actual values). There have been many other approaches to time-series modeling developed over the years. A few well-known paradigms, which have found their way into time-series analysis, include computational intelligence models like artificial neural networks (ANN) [10] and fuzzy logic [11].

The artificial neural network (ANN) is one of the popular techniques of machine learning which has been employed by researchers for developing a variety of models for time-series prediction. In principle, the ANN is used to learn complex non-linear functions and is primarily applied to classification problems [12] and regression problems [13]. In quite a few works, the time-series has been modeled by employing a regression neural network which takes $k$- time-lagged values $c(t-k),c(t-k+1),\ldots,c(t-2),c(t-1)$ of the time-series as input and produces the predicted value $c(t)$ of the time-series as output. A survey of early approaches pursued in this direction can be found in [14]. With time, more sophisticated neural models were developed to increase the accuracy of prediction on various time-series. Some of the works worth special mention are as follows.

An ensemble of ARIMA and ANN was proposed in [10] to combine the advantage of the linear modeling of ARIMA and the non-linear approximations of the ANN for better forecasts. A specialized ANN architecture was developed in [15] to improve the prediction accuracy on seasonal time-series, where the number of input and output neurons of the network is optimally chosen based on the duration of seasonality in the time-series. Yet another interesting work was proposed in [16], where the authors designed a recursive probabilistic extension of the well-known Levenberg-Marquardt algorithm to train recurrent neural networks for time-series modeling. Other notable works in this field include the use of recurrent neural network ensembles [17], hybrid Elman-NARX neural networks [18] and interval type-2 induced fuzzy back-propagation neural networks [19].

The advent of fuzzy set theory in the field of time-series analysis came with the works of Song and Chissom [20–22], where they formally provided the definition of a fuzzy time-series. A given time-series is fuzzified primarily by three steps. Firstly, the dynamic range of the time-series is divided into a number of equal or non-equal sized intervals called partitions, where each partition is a continuous section of the overall range of the series. Secondly, the dynamic range of the series is defined as the universe of discourse and each partition obtained from the first step is represented by a fuzzy set with a corresponding fuzzy membership function defined on the universe. Finally, each data point in the time-series is assigned a linguistic value based on the partition to which it (the data point) belongs with the highest membership. Multiple types of fuzzy implication rules including first-order rules having a single antecedent and a single consequent, and higher-order rules having multiple antecedents and a single consequent can be derived from the sequence of fuzzified time-series data points. These fuzzy implication rules can then be utilized for the task of prediction. A large number of interesting works [23–34] on fuzzy time-series exist in the literature, many of which use advanced models incorporating first and higher order fuzzy implication relations [23, 24], heuristic functions to improve fuzzy time-series prediction [25], and multi-factor fuzzy time-series models [26].

In the present work, we propose two ensemble neural network models where a set of regression neural networks is used to predict the future value of a time-series. For each pair of consecutive time-series data points $c(t-1)$ and $c(t)$, we construct a first-order transition rule of the form $P_{t-1} \rightarrow P_t$, where $P_{t-1}$ and $P_t$ represent the partitions to which the data points $c(t-1)$ and $c(t)$ belong respectively. Next, in the set $S$ of all such obtained transition rules, we identify the rules having the partition $P_i$ in the antecedent and include them in a separate set $S_i$. The idea is to group the transition rules into sets where all the rules contained within a set have the same antecedent. The probability of occurrence of a transition rule $P_i \rightarrow P_j$ is computed as the result of dividing the number of occurrences of the rule: $P_i \rightarrow P_j$ by the total number of occurrences of all rules: $P_i \rightarrow P_k, \forall k$.

In case the current time point falls in partition $P_i$, to predict the next possible partition, we need to fire the exhaustive set of rules containing $P_i$ in the antecedent. This apparently is possible, if all these concurrently fire-able rules are realized on different units of hardware/software. In the present settings, we planned to realize them on neural networks. Thus for $n$ concurrently fire-able rules having a common antecedent $P_i$, we require $n$ neural networks. The question that automatically is raised: should we realize a single rule on a neural net? This, of course is un-economic. We, therefore, plan to realize a set of rules, each with a distinct partition in the antecedent, on a neural network, thereby ensuring that no two or more rules from the set can fire concurrently.

Any traditional supervised neural network would suffice for our present purpose. We began our study with simple feed-forward neural nets pre-trained with the back-propagation algorithm. Here the antecedent and the consequent of the concurrently fire-able rules are used as input and output instances of the training patterns. Thus for training with $k$ rules we require a set of $k$ input-output training instances/patterns.

It is thus apparent that a neural network learns a subset of the set of all rules where each rule in the subset contains distinct partitions in the antecedent.

However, it may not realize the exhaustive set of rules with all possible partitions in the antecedent. In other words, consider a neural net realizing three rules with partitions $P_2, P_6$ and $P_7$ in the antecedents of these three rules. The entire set of rules may however include all the partitions $P_1$ through $P_{10}$ in the antecedent. Naturally, when the current partition is $P_3$, we would not use this neural net as $P_3$ does not appear in the antecedent of any of the rules realized by the neural network. This calls for a pre-selector that selects the right neural network depending on the occurrence of the partition corresponding to the current data point in the time-series as antecedent of one rule realized by the network. We here used a Radial Basis Function (RBF) [35] neural network to serve the purpose. The RBF network contains a hidden layer neuron corresponding to each distinct antecedent on which the neural network concerned is pre-trained. The neural network is triggered only when at least one of the hidden neurons in the pre-selector RBF is fired indicating that the current input partition occurs as the antecedent in a transition rule realized by the neural net. This ensures that only those networks which have been trained on rules containing the current partition as antecedent are triggered for prediction.

Although a time-series partition describes a bounded interval of a parameter (like temperature), it is generally represented by an approximate single value, such as the center of the interval. Such representation introduces an approximation error for all data points lying in the interval. One approach to overcome the creeping of approximation error is to describe each partition by a fuzzy set that takes into account, the distance between each data point and its respective partition-center while assigning memberships to the data points. These membership values indicate the degree of belongingness of the data points in a given partition. In the previous approach, we train each neural network in the ensemble with transition rules where the antecedents and consequents represent mid-point or center values of partitions. As the fuzzy representation for a partition, explained above, is relatively better than its mid-point representation, we attempt to incorporate the fuzzy representation in our neural network ensemble model. This can be done by fuzzifying the first-order transition rules where the antecedent of each rule is replaced by a membership function. The input instances in this case, for a neural net in the ensemble are the fuzzy membership values of the current time-series data point in every partition. Further, considering the effect of each partition in the antecedent of a transition rule eliminates the requirement of a pre-selector logic.

Experiments have been carried out to check the efficacy of the proposed models on real life chaotic time-series like the Sunspot [36] for comparison with existing neural network based time-series models. Also, the proposed models have also been applied on the Taiwan Stock Exchange Index (TAIEX) [37] economic close price time-series for the period 1990–2004 for comparison with existing fuzzy time-series models. The experiments indicate that the proposed models outperform the best among the existing models by a relatively large margin.

The remainder of this chapter is organized as follows. Section 5.2 provides a basic overview on fuzzy sets, the back-propagation training algorithm for neural networks and radial basis function networks. Section 5.3 discusses the first proposed model based on discrete first-order transition rules. Section 5.4 deals with the second proposed model applying fuzzy partitioning schemes for training. Section 5.5 contains the experiments carried out on various time-series as well as comparisons with other existing models. Conclusions are listed in Sect. 5.6.

## 5.2 Preliminaries

In this section, we discuss on some of the well-known concepts required for understanding the rest of the chapter. Specifically, the definitions of fuzzy sets, membership functions and time-series partitioning are provided along with a brief description of the back-propagation training algorithm for the neural network as well as Radial Basis Function networks.

### 5.2.1 *Fuzzy Sets and Time-Series Partitioning*

A fuzzy set is a generalization of a conventional set, where each element belongs with a certain membership value lying in the range [0,1] unlike a conventional set where each element belongs with a membership value which is either 0 (does not belong to the set) or 1 (belongs to the set). A fuzzy set is formally defined in Definition 5.1.

**Definition 5.1** Given a set U as the "universe of discourse", a fuzzy set A on the universe U is defined as a set of 2-tuples as shown below

$$A = \{(x, \mu_A(x)) | x \in U\} \tag{5.1}$$

where x is an element of the universe U and $\mu_A(x) \in [0, 1]$denotes the membership value of the element x in the fuzzy set A. If the universe of discourse U is continuous and infinite, it is not possible to define a set of discrete 2-tuples as given in Eq. (5.1) for the fuzzy set A. In such cases, a fuzzy membership function $\mu_A : U \to [0, 1]$ is defined to map each element in the universe of discourse to its corresponding membership value in the fuzzy set A.

**Definition 5.2** A time-series is a discrete sequence of values sampled from a temporally varying measurable entity like atmospheric temperature, rainfall, population growth and the like. A time-series can be considered as a vector of length n as shown below:

$$\begin{aligned}\vec{C} &= [c(t-(n-1)k), c(t-(n-2)k), \ldots, c(t-2), c(t-1), c(t)] \\ &= [c_1, c_2, c_3, \ldots, c_n]\end{aligned} \tag{5.2}$$

where $c_i = c(t-(n-i)k)$ represents the ith sample of the series and k is the sampling interval.

**Definition 5.3** Given a time-series $\vec{C}$, let $c_{max}$ and $c_{min}$ represent the maximum and minimum values of the time-series respectively. We define partitioning as the act of dividing the range R = [$c_{min}$, $c_{max}$], into h non-overlapping contiguous intervals $P_1, P_2, \ldots, P_h$ such that the following two conditions jointly hold:

$$P_i \cap P_j = \emptyset, \quad \forall i \neq j,\ i,j \in \{1,2,\ldots,h\} \tag{5.3}$$

$$\bigcup_{i=1}^{h} P_i = R, \quad \text{for integer } i. \tag{5.4}$$

Let a partition $P_i = [P_i^-, P_i^+]$ be defined as a bounded set where $P_i^-$ and $P_i^+$ represent the lower and upper bounds of the partition respectively. Clearly, a time-series data point c(t) belongs to the partition $P_i$ if and only if $P_i^- \leq c(t) \leq P_i^+$.

**Definition 5.4** Let $c(t)$ and $c(t+1)$ denote two consecutive data points in a time-series $\vec{C}$ and let $P_i$ and $P_j$ be the partitions to which these data points belong. Then we denote $P_i \rightarrow P_j$ as a first-order transition rule that exists in the time-series $\vec{C}$, with the antecedent $P_i$ and consequent $P_j$. The following example demonstrates the first-order transition rule in a time-series.

*Example 5.1* Let a given time-series $\vec{C} = [3,5,10,8,3,1,9,2]$, be divided into five partitions $P_1 = [0,2), P_2 = [2,4), P_3 = [4,6), P_4 = [6,8), P_5 = [8,10]$. Clearly, the sequence of partitions corresponding to each data point in the time-series is $\vec{P} = [P_2, P_3, P_5, P_5, P_2, P_1, P_5, P_2]$. From the sequence $\vec{P}$, the first-order transition rules obtained, are

$$P_2 \rightarrow P_3, P_3 \rightarrow P_5, P_5 \rightarrow P_5, P_5 \rightarrow P_2, P_2 \rightarrow P_1, P_1 \rightarrow P_5, P_5 \rightarrow P_2.$$

### 5.2.2 Back-Propagation Algorithm

One of the most well-known training algorithms employed to optimize the weights of a neural network is the back-propagation algorithm. The basic principle behind the back-propagation algorithm is to minimize a loss function by iterative modification of the weights of the network over all the training examples. The algorithm uses gradient descent learning on an error (energy) function of weights, where the error function is designed to minimize the Euclidean norm of two vectors: targeted output vector and computed output vector. Here the vectors are defined with respect to the output signals corresponding to the neurons in the last layer. The algorithm begins by randomly initializing the weights of a neural network and has two primary steps which are iteratively carried out over all the training examples until the network performance is satisfactory:

Step 1. *Forward Propagation*: In this step, a training example is input in order to obtain the output activation values from the network and compute the loss or error metric $e$ of the network output with respect to the desired output.

Step 2. *Backward Propagation and Weight Update*: This step is used to compute the gradient $\frac{\partial e}{\partial w}$ of the error (or loss) with respect to each and every weight $w$ in the network. Furthermore, for a weight $w$, the negative of the gradient $\frac{\partial e}{\partial w}$ represents the direction of decreasing loss. Hence, the weight $w$ is updated as follows:

$$w_i \leftarrow w_{i-1} - \alpha\left(\frac{\partial e_{i-1}}{\partial w_{i-1}}\right) \tag{5.5}$$

where $w_i$ and $w_{i-1}$ represent the weight at the $ith$ and $(i-1)th$ iterations respectively, $e_{i-1}$ is the error at the $(i-1)th$ iteration, $\frac{\partial e_{i-1}}{\partial w_{i-1}}$ is the value of the gradient and $\alpha$ denotes the learning rate of the network.

### 5.2.3 Radial Basis Function (RBF) Networks

A Radial Basis Function (RBF) network is a 3-layer (input, hidden and output) neural network which uses non-linear radial basis functions as activation functions in its hidden layer. The output of the network is generally a linear combination of the hidden layer activation values. Let the input to the network be represented as a vector $\vec{X} = (x_1, x_2, \ldots, x_k)$ and let there be $h$ hidden layer neurons in the network. The output of the RBF network is then given as:

$$\phi(\vec{X}) = \sum_{i=1}^{h} a_i \rho\left(\|\vec{X} - \vec{C}_i\|\right) \tag{5.6}$$

where $\vec{C}_i$ represents the central vector of the $ith$ hidden layer neuron and the norm $\|\vec{X} - \vec{C}_i\|$, is a distance measure like the Euclidean distance between the vectors $\vec{X}$ and $\vec{C}_i$. Intuitively, each hidden layer neuron outputs the similarity between the input vector and its own central vector. Hence, a commonly used radial basis function is the Gaussian function as given below

$$\rho\left(\|\vec{X} - \vec{C}_i\|\right) = e^{-\beta\left(\|\vec{X} - \vec{C}_i\|\right)^2} \tag{5.7}$$

where $\beta$ is a positive constant. Due to the dependence of the activation function on the distance between the input vector and a hidden neuron's central vector, the function is radially symmetric about the central vector of the neuron. In this chapter, we use RBF neurons as pre-selectors to determine the neural networks to be triggered based on the transition rules on which the networks have been pre-trained.

## 5.3 First-Order Transition Rule Based NN Model

In this section, we propose a neural network ensemble model trained on first-order transition rules obtained from a given time-series. Let $S$ be the set of all first-order transition rules and let $S_i$ denote the set of rules having the partition $P_i$ in the antecedent. Further, let $p(P_j/P_i)$ indicate the probability of occurrence of $P_j$ as the next partition, given that $P_i$ is the current partition. Clearly, the value of $p(P_j/P_i)$ is computed as given below:

$$p(P_j/P_i) = \frac{count(P_i \to P_j)}{\sum\limits_{\forall k} count(P_i \to P_k)} \quad (5.8)$$

where $count(P_i \to P_j)$ represents the number of occurrences of the transition rule $P_i \to P_j$. The set $S_i$ can then be represented as follows:

$$S_i = \{P_i \to P_k | p(P_k/P_i) > 0\}. \quad (5.9)$$

Given the current partition $P_i$, in order to make a prediction for the next partition, it is desirable to consider the weighted contribution of all the transition rules having $P_i$ in the antecedent. This can be best achieved by designing a model that allows for the simultaneous firing of all the rules in the set $S_i$. In order to realize such a model, we use an ensemble of feed-forward neural networks which satisfies the following conditions:

*Condition 1*: Given the current input partition $P_i$, each neural network in the ensemble can fire at most a single transition rule having $P_i$ in the antecedent.

*Condition 2*: All the transition rules having partition $P_i$ in the antecedent must be fired by the ensemble. In other words, all the rules contained in the set $S_i$ must be fired.

*Condition 3*: No two neural networks in the ensemble can fire the same transition rule, i.e., for any two rules $P_i \to P_a$ and $P_i \to P_b$ fired simultaneously, $P_a \neq P_b$.

The above mentioned conditions can be satisfied by imposing certain constraints on the training sets of the neural networks. A few observations that are evident from the above conditions are given as follows:

**Theorem 5.1** *For any neural network $NN_i$ in the ensemble, its training set $T_i$ should not contain multiple transition rules having the same partition in the antecedent.*

*Proof* We prove this theorem by the method of contradiction. Let the training set $T_i$ contain two transition rules $P_a \rightarrow P_b$ and $P_a \rightarrow P_c$ having the same partition $P_a$ in the antecedent. If the current input partition is $P_a$,following Condition 1, any one of the rules $P_a \rightarrow P_b$ or $P_a \rightarrow P_c$ will be fired by the network $NN_i$. Without any loss of generality, let us arbitrarily assume that the fired rule is $P_a \rightarrow P_b$. Hence, the rule $P_a \rightarrow P_c$ is not fired. By Condition 2, we can say that the rule $P_a \rightarrow P_c$ is fired by some other neural network $NN_j$ in the ensemble. However, the decision to fire $P_a \rightarrow P_b$ is completely arbitrary and the rule $P_a \rightarrow P_b$ could have been fired as well, violating Condition 3. Hence, we arrive at a contradiction and our initial assumption was incorrect. The training set $T_i$ does not contain multiple rules having the same partition in the antecedent. □

**Theorem 5.2** *For any two neural networks $NN_i$ and $NN_j$ in the ensemble having training sets $T_i$ and $T_j$ respectively, the training sets are disjoint, i.e., $T_i \cap T_j = \emptyset$.*

*Proof* We prove this theorem by the method of contradiction. Let us assume that the training sets $T_i$ and $T_j$of the two neural networks $NN_i$ and $NN_j$ contain the same transition rule $P_a \rightarrow P_b$. Hence, when the current input partition is $P_a$,both the networks $NN_i$ and $NN_j$ fire the same rule $P_a \rightarrow P_b$ as, from Theorem 5.1, they have no other rule in their training sets with $P_a$ in the antecedent. This however, violates Condition 3 mentioned above thereby contradicting our initial assumption that the two training sets have a transition rule which is common. Thus, all the training sets are completely disjoint with respect to one another. □

Bearing in mind the above conditions, we group the set *S*o transition rules into training sets using the following steps:

**Step 1.** The set *S* is divided into groups or subsets of transition rules, where all the rules in a group have the same partition in the antecedent. Let the subset of rules having $P_i$ in the antecedent be denoted by $S_i$ following Eq. (5.8). Clearly, the number of such subsets is equal to the total number of distinct partitions occurring in the antecedent of the transition rules.

**Step 2.** The transition rules in each subset are ordered (following any arbitrary order). Now, the training set for the first neural net in the ensemble is constructed by collecting the transition rules which occur in the first position in each subset, the training set for the second neural net is constructed by collecting the transition rules occurring in the second position in each subset and so on. The algorithm for the construction of training sets is formally presented in Pseudo Code 5.1. Also, for better comprehension, an illustrative example of the same is provided in Example 5.2.

**Pseudo Code 5.1: Training Set Construction Algorithm**

**Input:** A set S of all first-order transition rules obtained from a time-series.

**Output:** A sequence of training sets $T_1, T_2, \ldots, T_v$ where the *ith* training set $T_i$ contains the transition rules on which the *ith* neural network $NN_i$ is trained.

BEGIN

Let k be the number of distinct partitions occurring in the antecedent of the transition rules in S;

Group the transition-rules in S into k arrays $S_1, S_2, \ldots, S_k$ such that the array $S_i,\ i \in \{1, 2, \ldots, k\}$ contains only the rules having partition $P_i$ in the antecedent;

$j \leftarrow 1$;//denotes the training set number

WHILE construction of training sets is not complete, DO

BEGIN

Initialize $T_j \leftarrow \emptyset$; //$T_j$ is initially an empty set

FOR each array $S_i,\ i \in \{1, 2, \ldots, k\}$ DO

BEGIN

Let $P_x \to P_y$ be the first rule (if any) in the array $S_i$;

$T_j \leftarrow T_j \cup \{P_x \to P_y\}$;//The rule is added to the training set

Remove the rule $P_x \to P_y$from the array $S_i$;

END FOR;

IF all the arrays $S_1, S_2, \ldots, S_k$ are empty, THEN

Break from WHILE loop; //Training set construction is complete

END IF;

$j \leftarrow j + 1$;// Training set $T_j$ has been constructed, moving to construction of $T_{j+1}$

END WHILE;

END

*Example 5.2* Let the transition rules obtained from the time-series be represented as the set $S$ as given below:

$$S = \{P_1 \to P_1, P_1 \to P_2, P_2 \to P_1, P_2 \to P_2, P_2 \to P_3, P_3 \to P_1, P_3 \to P_3\}. \quad (5.10)$$

First, we group the rules in $S$ according to the common antecedents. Hence, we obtain three ordered subsets of $S$, i.e., $S_1 = \{P_1 \to P_1, P_1 \to P_2\}$, $S_2 = \{P_2 \to P_1, P_2 \to P_2, P_2 \to P_3\}$ and $S_3 = \{P_3 \to P_1, P_3 \to P_3\}$. For the construction of the first training set $T_1$, we collect the first transition rules in each set $S_1, S_2$ and $S_3$. Hence, the training set $T_1$ is constructed as follows:

$$T_1 = \{P_1 \rightarrow P_1, P_2 \rightarrow P_1, P_3 \rightarrow P_1\}. \quad (5.11)$$

The training sets $T_2$ and $T_3$ contain the second and third transition rules in $S_1, S_2$ and $S_3$ and are given as follows:

$$T_2 = \{P_1 \rightarrow P_2, P_2 \rightarrow P_2, P_3 \rightarrow P_3\} \quad (5.12)$$

$$T_3 = \{P_2 \rightarrow P_3\}. \quad (5.13)$$

It should be noted that the training set $T_3$ contains only a single transition rule as there are no rules left in the sets $S_1$ and $S_3$ after the construction of the training sets $T_1$ and $T_2$. □

Having constructed the training sets, we are ready to train the neural nets in the ensemble using the back-propagation algorithm. It should be noted that the training sets contain transition rules, where both the antecedent and consequent of each rule represents a partition label like $P_1$. However, our current task demands a computational model whose expected input is the current time-series data point and the desired output is the next or future time-series data point, both of which are real numbers. This requires us to use regression neural networks in the ensemble whose input and output values are both real numbers. Hence, in order to appropriately convert the training set to fit our needs, we replace each partition label $P_i$ with its corresponding partition mid-value $m_i$. With this modification to the training sets in place, we are able to train the neural networks in the ensemble.

During the prediction phase, let the partition containing the current time-series data point (input to the system) be $P_i$. It may so happen, that a neural network in the ensemble has not been trained on a rule having $P_i$ in the antecedent. Clearly, it is undesirable to use such a network for prediction as that may lead to erroneous results. In order to deal with this problem, we use a set of RBF neurons as pre-selectors for triggering appropriate neural networks in the ensemble based on the given input. Let the total number of partitions be $h$. We then construct $h$ RBF neurons, one for each partition. The output activation value $\rho_i$ for the *ith* RBF neuron is given as follows:

$$\rho_i = e^{-\beta\left(\left|x^2 - m_i^2\right|\right)} \quad (5.14)$$

where $x$ is the input to the RBF neuron, $m_i$ is the mid-point value of the *ith* partition $P_i$ and $\beta$ is an empirically chosen constant. Intuitively, the *ith* RBF neuron outputs the similarity between the input and the mid-value of the *ith* partition $P_i$. Given the current input partition $P_c$, our objective is to trigger only those neural nets which have been trained on at least one rule having $P_c$ in the antecedent. In other words, a neural network $NN_j$ is enabled, if the current input partition $P_c$ is present in the antecedent of any one of the training set rules for $NN_j$. Hence, the enable signal for $NN_j$ is obtained by a logical OR operation of the outputs of the RBF neurons which correspond to the partitions occurring in the antecedent of the training set rules for $NN_j$. The operands for logical OR being binary, we have to apply a simple

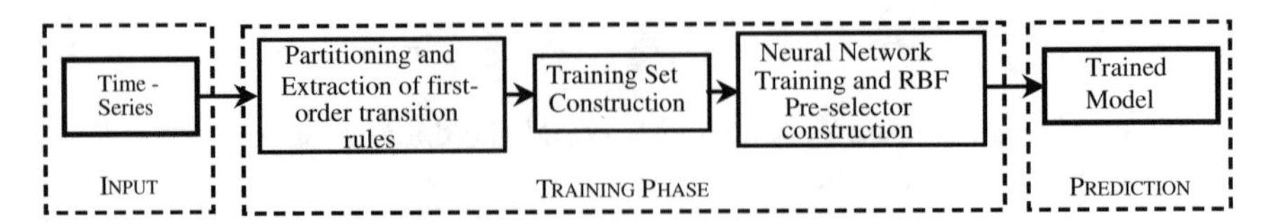


**Fig. 5.1** Schematic block diagram of the steps proposed in the neural network ensemble model

thresholding technique to the outputs of the RBF neurons to convert them to binary values. An illustrative example of the pre-selection technique is given in Example 5.3.

*Example 5.3* Let us consider the transition rules given in Eq. (5.10) of Example 5.2. There are three partitions in the rules, i.e., $P_1, P_2$ and $P_3$. We construct three RBF neurons $RBF_1, RBF_2$ and $RBF_3$ corresponding to the partitions $P_1, P_2$ and $P_3$ respectively. Now, the training set $T_1 = \{P_1 \rightarrow P_1, P_2 \rightarrow P_1, P_3 \rightarrow P_1\}$ for the neural net $NN_1$ contains all the three partitions $P_1, P_2$ and $P_3$ in the antecedents of its transition rules. Hence, the enable signal for $NN_1$ is obtained by a logical OR of the outputs of $RBF_1, RBF_2$ and $RBF_3$. Similarly, the neural net $NN_2$ is also enabled by a logical OR of the outputs of $RBF_1, RBF_2$ and $RBF_3$. However, the training set $T_3 = \{P_2 \rightarrow P_3\}$ for the neural net $NN_3$ contains only the partition $P_2$ in the antecedent. Hence, the enable signal for $NN_3$ can be directly obtained from the output of $RBF_2$. □

The final prediction $c'(t+1)$ for the next time-series data point, given the current data point $c(t)$, is computed as the weighted sum of the individual outputs obtained from the neural nets which are triggered by the pre-selector RBF neurons. Let there be $v$ selected neural nets and let their respective outputs be $o_1, o_2, \ldots, o_v$ lying in the partitions $P_{o1}, P_{o2}, \ldots, P_{ov}$ respectively. The final output of the ensemble is then given as follows:

$$c'(t+1) = \sum_{i=1}^{v} (o_i \times p(P_{oi}/P_c)) \tag{5.15}$$

where $P_c$ is the partition containing the current time-series data point $c(t)$ and $p(P_{oi}/P_c)$ is the probability of occurrence of $P_{oi}$ as the next partition, given that the current partition is $P_c$. A schematic block diagram of the steps in our proposed approach is illustrated in Fig. 5.1. The architecture of the ensemble corresponding to the set of transition rules given in Example 5.2 is shown in Fig. 5.2. It should be noted that in our chapter, we have designed each neural network in the ensemble with a single hidden layer having 10 hidden neurons as shown in Fig. 5.3.

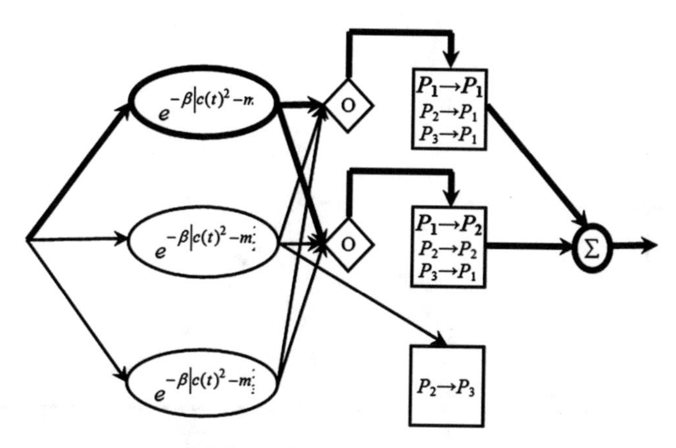


**Fig. 5.2** Architecture of the ensemble corresponding to Example 5.1, where $m_1$, $m_2$ and $m_3$ are the mid-values of the partitions $P_1$, $P_2$ and $P_3$ respectively, the given input lies in partition $P_1$, hence, $RBF_1$ is fired and the corresponding rules in $NN_1$ and $NN_2$ are fired. The fired paths and rules are shown in *bold*

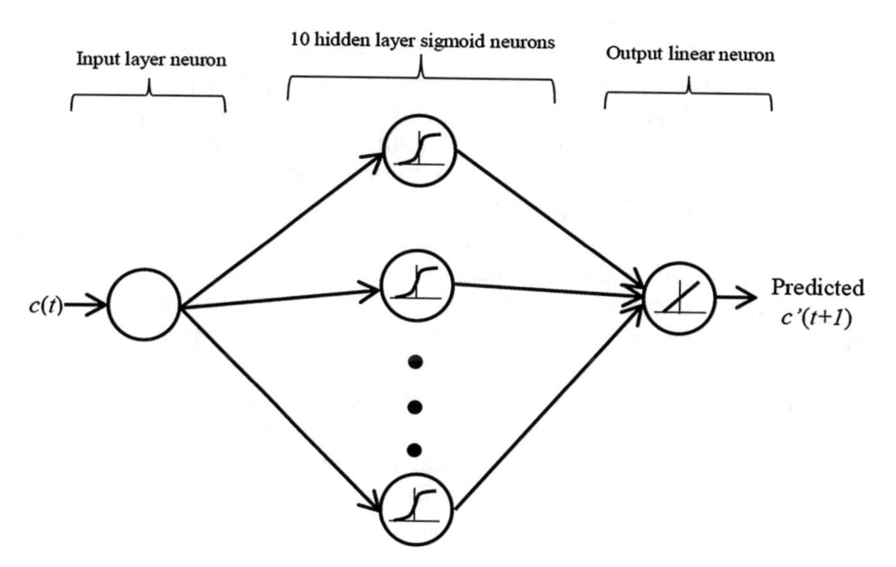


**Fig. 5.3** Neural network architecture for the first-order transition rule based NN model, c(t) is the input time-series value and c'(t + 1) is the predicted future value by the neural network

## 5.4 Fuzzy Rule Based NN Model

In this section, we propose a variation in the training scheme of the neural nets from the previous model. It should be noted that a time-series data point cannot be wholly represented by a partition. A partition is a certain continuous portion of the

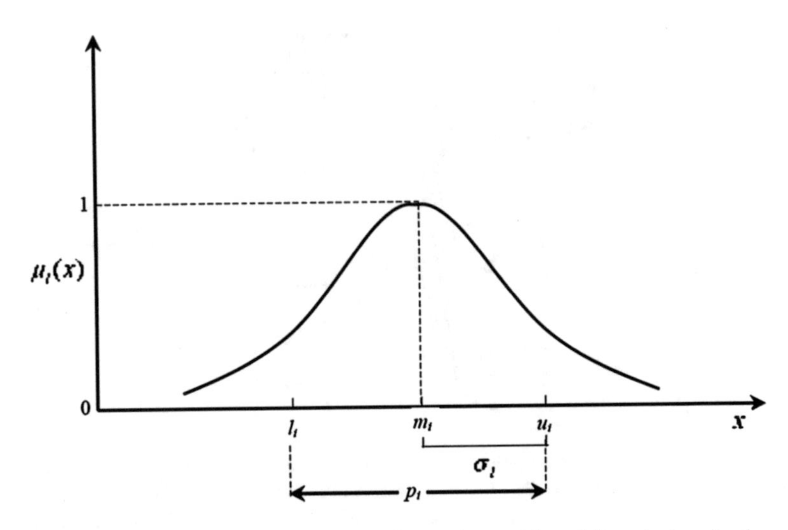


**Fig. 5.4** Membership function μi(x) corresponding to the partition *pi*, *li*, and *ui* are the *lower* and *upper bounds* of the partition *pi* and *mi* is the corresponding mid-point of the partition

dynamic range of the time-series. The best way to represent a partition is to consider its mid-point. Hence, by approximating a data point lying in a partition with its corresponding mid-point value, we intentionally delve into some approximation error which grows with the width of the partition itself. One way to avoid this error is to consider each partition as a fuzzy set, with the dynamic range of the time-series representing the universe of discourse. Clearly, each partition is then associated with a judiciously chosen membership function and every time-series data point will have some membership value lying in the set [0,1], for a partition.

We want to design a membership function with the peak corresponding to the center of the partition and a gradual fall-off on either side of the peak with a value almost close to 0 towards the bounds of the partition. The membership value need not however necessarily be 0 at the boundaries of the partition. One typical function which satisfies this requirement is the Gaussian membership function as shown in Fig. 5.4. We use these membership values to train the neural networks in this model. The advantage of this approach is the exploitation of the inherent fuzziness involved in the task of assigning a partition to a time-series data point and using it for better prediction results.

Let the *ith* partition be $p_i = [l_i, u_i]$,and let the mid-point of the partition be $m_i$. We define the standard deviation for the Gaussian membership function as $\sigma_i = (u_i - m_i) = (m_i - l_i)$. With the above terms defined, the membership function corresponding to the *ith* partition is given in Eq. (5.16) and illustrated in Fig. 5.4.

$$\mu_i(x) = e^{-\frac{(x-m_i)^2}{2\sigma_i^2}} \tag{5.16}$$

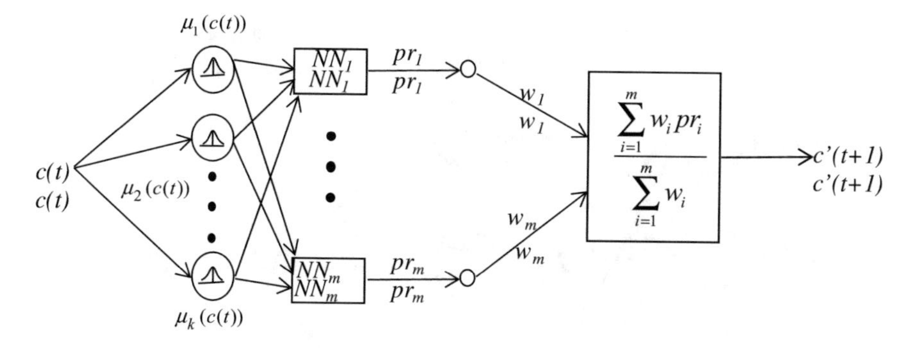


**Fig. 5.5** Schematic diagram of the fuzzy rule based neural network model (considering k partitions), c(t) is the input time-series data point, $\mu_i(c(t))$ denotes the membership value of c(t) in the ith partition, $NN_i$ is the ith neural network producing the prediction $pr_i$ attached with weight $w_i$ and c'(t+1) represents the predicted time-series value

Let there be $k$ partitions in the time-series. For each first-order transition rule of the form $p_i \rightarrow p_j$, ($p_i$ and $p_j$ represent partitions) a new rule of the form $(\mu_1(m_i), \mu_2(m_i), \ldots, \mu_k(m_i)) \rightarrow m_j$ is created, where $m_i$ and $m_j$ are the mid-points of partitions $p_i$ and $p_j$ respectively and $\mu_l(m_i)$, $l \in \{1, 2, \ldots, k\}$, are the membership values of $m_i$ in each partition of the time-series. The modified rules are used to train the neural network. A schematic of the proposed model is shown in Fig. 5.5. Detailed steps of the approach are given as follows.

Step 1. *Partitioning*: This is the same as the discrete rule based model discussed in Sect. 5.3.

Step 2. *First-order transition rule construction*: This is the same as the discrete rule based model discussed in Sect. 5.3.

Step 3. *Neural network training*: As discussed in Sect. 5.3, let the training set $T$ obtained for a neural network be defined as $T = \{(p_{iq}, p_{jq}) | q \in \{1, 2, \ldots, r\}\}$, i.e., a set of $r$ 2-tuples where the first and second elements of each tuple represent respectively, the antecedent and consequent of a first-order transition rule. For the proposed extension, the training set $T$ is modified as follows.

$$T = \{((\mu_1(m_{iq}), \mu_2(m_{iq}), \ldots, \mu_k(m_{iq})), m_{jq}) | q \in \{1, 2, \ldots, r\}\} \quad (5.17)$$

where $m_{iq}$ and $m_{jq}$ are the mid-points of the partitions $p_{iq}$ and $p_{jq}$ respectively and $\mu_l(x)$ is the membership value of $x$ in the fuzzy set, representing the degree of closeness of x with the centre of the $lth$ partition. The architecture of the neural networks is shown in Fig. 5.6.

Step 4. *Predicting the next time-series data point*: This is the same as the discrete rule based model discussed in Sect. 5.3.

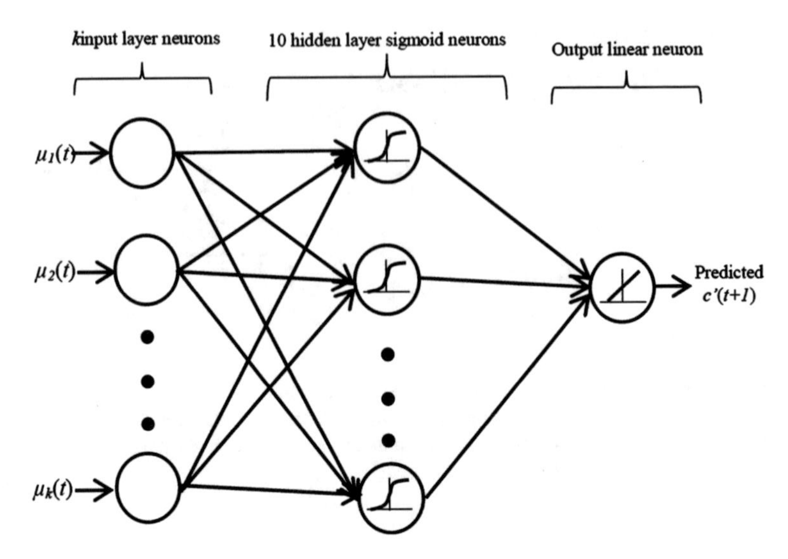


**Fig. 5.6** Neural network architecture for the fuzzy transition rule based model, input layer neurons correspond to the membership value of a time-series data point c(t) in each of the k partitions

## 5.5 Experiments and Results

In this section, we present the results obtained from two experiments carried out to test the prediction efficacy of the proposed models. In the first experiment, we apply our models to the well-known Sunspot time-series data from November 1834 to June 2001 for comparison with existing neural time-series prediction models in the literature. In the second experiment, we use the TAIEX economic close-price time-series for prediction in the field of trading and investment as well as for comparison with existing fuzzy time-series models. Both the experiments are performed using MATLAB on an Intel Core i7 processor with a clock speed of 2.3 GHz and 8 GB of RAM. The details of each experiment are provided in the following sub-sections.

### *5.5.1 Experiment 1: Sunspot Time-Series Prediction*

The sunspot time-series data is used to record solar activity and is a classic example of a real-life chaotic time-series. Due to the effect of solar activity on various aspects of human life like the climate and weather, prediction of the sunspot time-series has become an important and critical challenge for many researchers. In this chapter, we use the smoothed sunspot time-series data from the World Data Center for sunspot index. The time-series from the period of November 1834 to

**Table 5.1** Transition matrix obtained for the first-order transition rules from sunspot training phase time-series

| To Partition From partition | $P_1$ | $P_2$ | $P_3$ | $P_4$ | $P_5$ | $P_6$ | $P_7$ | $P_8$ | $P_9$ | $P_{10}$ | $P_{11}$ | $P_{12}$ | $P_{13}$ | $P_{14}$ | $P_{15}$ | $P_{16}$ | $P_{17}$ | $P_{18}$ | $P_{19}$ | $P_{20}$ |
|---|---|---|---|---|---|---|---|---|---|---|---|---|---|---|---|---|---|---|---|---|
| $P_1$ | 0.7561 | 0.2098 | 0.0293 | 0.0049 | 0 | 0 | 0 | 0 | 0 | 0 | 0 | 0 | 0 | 0 | 0 | 0 | 0 | 0 | 0 | 0 |
| $P_2$ | 0.3909 | 0.3182 | 0.2 | 0.0727 | 0.0091 | 0.0091 | 0 | 0 | 0 | 0 | 0 | 0 | 0 | 0 | 0 | 0 | 0 | 0 | 0 | 0 |
| $P_3$ | 0.0522 | 0.2708 | 0.3333 | 0.1979 | 0.0729 | 0.0625 | 0 | 0.0104 | 0 | 0 | 0 | 0 | 0 | 0 | 0 | 0 | 0 | 0 | 0 | 0 |
| $P_4$ | 0.0233 | 0.0465 | 0.1860 | 0.2558 | 0.2209 | 0.1163 | 0.1047 | 0.0233 | 0.0116 | 0.0116 | 0 | 0 | 0 | 0 | 0 | 0 | 0 | 0 | 0 | 0 |
| $P_5$ | 0 | 0.0098 | 0.0882 | 0.2255 | 0.2549 | 0.2255 | 0.1471 | 0.0392 | 0 | 0 | 0 | 0 | 0 | 0.0098 | 0 | 0 | 0 | 0 | 0 | 0 |
| $P_6$ | 0 | 0.0108 | 0.0645 | 0.0860 | 0.2581 | 0.2366 | 0.1398 | 0.1398 | 0.0215 | 0.0430 | 0 | 0 | 0 | 0 | 0 | 0 | 0 | 0 | 0 | 0 |
| $P_7$ | 0 | 0 | 0.0270 | 0.0405 | 0.1892 | 0.2162 | 0.2568 | 0.1351 | 0.1081 | 0.0135 | 0.0135 | 0 | 0 | 0 | 0 | 0 | 0 | 0 | 0 | 0 |
| $P_8$ | 0 | 0 | 0.0164 | 0.0328 | 0.1475 | 0.1148 | 0.1639 | 0.1803 | 0.1148 | 0.0984 | 0.0492 | 0.0492 | 0.0164 | 0.0164 | 0 | 0 | 0 | 0 | 0 | 0 |
| $P_9$ | 0 | 0 | 0.0233 | 0 | 0.0233 | 0.0233 | 0.1163 | 0.1860 | 0.2093 | 0.2326 | 0.0698 | 0.0930 | 0.0233 | 0 | 0 | 0 | 0 | 0 | 0 | 0 |
| $P_{10}$ | 0 | 0 | 0 | 0 | 0.0244 | 0.0976 | 0.0244 | 0.1463 | 0.1463 | 0.2927 | 0.1707 | 0.0244 | 0.0488 | 0.0244 | 0 | 0 | 0 | 0 | 0 | 0 |
| $P_{11}$ | 0 | 0 | 0 | 0 | 0 | 0.1034 | 0.0345 | 0.1379 | 0.1034 | 0.1724 | 0.2069 | 0.1034 | 0 | 0 | 0.1379 | 0 | 0 | 0 | 0 | 0 |
| $P_{12}$ | 0 | 0 | 0 | 0 | 0 | 0 | 0.1250 | 0.0625 | 0.1250 | 0.0625 | 0.0625 | 0.1250 | 0.25 | 0.0625 | 0.0625 | 0 | 0 | 0 | 0 | 0.0625 |
| $P_{13}$ | 0 | 0 | 0 | 0 | 0 | 0 | 0 | 0 | 0.2222 | 0.0556 | 0.2778 | 0.0556 | 0.2222 | 0.1111 | 0.0555 | 0 | 0 | 0 | 0 | 0 |
| $P_{14}$ | 0 | 0 | 0 | 0 | 0 | 0 | 0 | 0.1111 | 0.1111 | 0 | 0.2222 | 0.1111 | 0.1111 | 0.1111 | 0.1111 | 0.1111 | 0 | 0 | 0 | 0 |
| $P_{15}$ | 0 | 0 | 0 | 0 | 0 | 0 | 0 | 0 | 0 | 0 | 0.1250 | 0.1250 | 0.1250 | 0 | 0.1250 | 0.1250 | 0.3750 | 0 | 0 | 0 |
| $P_{16}$ | 0 | 0 | 0 | 0 | 0 | 0 | 0 | 0 | 0 | 0 | 0 | 0 | 0.5 | 0.5 | 0 | 0 | 0 | 0 | 0 | 0 |
| $P_{17}$ | 0 | 0 | 0 | 0 | 0 | 0 | 0 | 0 | 0 | 0 | 0 | 0 | 0.75 | 0.25 | 0 | 0 | 0 | 0 | 0 | 0 |
| $P_{18}$ | 0 | 0 | 0 | 0 | 0 | 0 | 0 | 0 | 0 | 0 | 0 | 0 | 0 | 0 | 0 | 0 | 1 | 0 | 0 | 0 |
| $P_{19}$ | 0 | 0 | 0 | 0 | 0 | 0 | 0 | 0 | 0 | 0 | 0 | 0 | 0 | 0 | 0 | 0 | 0 | 0 | 0 | 0 |
| $P_{20}$ | 0 | 0 | 0 | 0 | 0 | 0 | 0 | 0 | 0 | 0 | 0 | 0 | 0 | 0 | 0 | 0 | 0 | 1 | 0 | 0 |

June 2001 has 2000 data-points which is divided into two equal parts of 1000 points each. The first half is used for training the models and the second half is used for testing. The time-series is scaled to the range [0, 1]. The following steps are carried out for the experiment:

*Step 1. Partitioning, First-Order Rule Extraction And Neural Network Training*: In this step, we first partition the training phase time-series into 20 partitions. We experimentally choose the value 20 based on best prediction results. The first-order transition rules extracted from the partitioned time-series along with the probability of occurrence of each rule is given in the transition matrix shown in Table 5.1. It should be noted that the entry corresponding to the cell ($P_i$, $P_j$) contains the probability of occurrence of the first-order transition $P_i \rightarrow P_j$.

Following the extraction of first-order transition rules, they are grouped into training sets, each representing a mapping of distinct antecedents to consequents. The groups of transition rules used to train each neural network is shown in Table 5.2. It should be noted that groups with less than 6 transition rules are ignored for training purposes.

The bunched first-order transition rules are further processed to yield training sets for the two proposed neural network ensembles according to Sects. 5.3 and 5.4. The networks are trained and we use the trained ensembles of both the proposed models to make predictions on the test phase time-series.

*Step 2. Prediction on Test-Phase Time-Series*: In this step, we apply the trained models to make predictions on the test phase Sunspot time-series. Figs. 5.7 and 5.8 illustrate the predictions made by the first-order rule based NN model and the fuzzy rule based NN model respectively on the test phase sunspot series. In order to quantify the prediction accuracy, we use three well-known error metrics, i.e., the mean square error (MSE), the root mean square error (RMSE) and the normalized mean square error (NMSE). Let $c_{test}(t)$denote the value of the test-period time-series at the time-instant $t$ and let $c'(t)$ be the predicted time-series value for the same time-instant. The above mentioned error metrics can be defined by the following equations:

$$MSE = \frac{\sum_{t=1}^{N} (c_{test}(t) - c'(t))^2}{N} \tag{5.18}$$

$$RMSE = \sqrt{\frac{\sum_{t=1}^{N} (c_{test}(t) - c'(t))^2}{N}} \tag{5.19}$$

**Table 5.2** groups of first-order transition rules used to train each neural network ($NN_1$ to $NN_9$) for the sunspot time-series. $NN_{10}$ to $NN_{12}$ are not trained as the number of rules obtained for their respective training sets is below the minimum threshold of 6

| Antecedent | $NN_1$ | $NN_2$ | $NN_3$ | $NN_4$ | $NN_5$ | $NN_6$ | $NN_7$ | $NN_8$ | $NN_9$ | $NN_{10}$ | $NN_{11}$ | $NN_{12}$ |
|---|---|---|---|---|---|---|---|---|---|---|---|---|
| $P_1$ | $P_1 \to P_1$ | $P_1 \to P_2$ | $P_1 \to P_3$ | $P_1 \to P_4$ | | | | | | | | |
| $P_2$ | $P_2 \to P_1$ | $P_2 \to P_2$ | $P_2 \to P_3$ | $P_2 \to P_4$ | $P_2 \to P_5$ | $P_2 \to P_6$ | | | | | | |
| $P_3$ | $P_3 \to P_1$ | $P_3 \to P_2$ | $P_3 \to P_3$ | $P_3 \to P_4$ | $P_3 \to P_5$ | $P_3 \to P_6$ | $P_3 \to P_8$ | | | | | |
| $P_4$ | $P_4 \to P_1$ | $P_4 \to P_2$ | $P_4 \to P_3$ | $P_4 \to P_4$ | $P_4 \to P_5$ | $P_4 \to P_6$ | $P_4 \to P_7$ | $P_4 \to P_8$ | $P_4 \to P_9$ | $P_4 \to P_{10}$ | | |
| $P_5$ | $P_5 \to P_2$ | $P_5 \to P_3$ | $P_5 \to P_4$ | $P_5 \to P_5$ | $P_5 \to P_6$ | $P_5 \to P_7$ | $P_5 \to P_8$ | $P_5 \to P_{14}$ | | | | |
| $P_6$ | $P_6 \to P_2$ | $P_6 \to P_3$ | $P_6 \to P_4$ | $P_6 \to P_5$ | $P_6 \to P_6$ | $P_6 \to P_7$ | $P_6 \to P_8$ | $P_6 \to P_9$ | $P_6 \to P_{10}$ | | | |
| $P_7$ | $P_7 \to P_3$ | $P_7 \to P_4$ | $P_7 \to P_5$ | $P_7 \to P_6$ | $P_7 \to P_7$ | $P_7 \to P_8$ | $P_7 \to P_9$ | $P_7 \to P_{10}$ | $P_7 \to P_{11}$ | | | |
| $P_8$ | $P_8 \to P_3$ | $P_8 \to P_4$ | $P_8 \to P_5$ | $P_8 \to P_6$ | $P_8 \to P_7$ | $P_8 \to P_8$ | $P_8 \to P_9$ | $P_8 \to P_{10}$ | $P_8 \to P_{11}$ | $P_8 \to P_{12}$ | $P_8 \to P_{13}$ | $P_8 \to P_{14}$ |
| $P_9$ | $P_9 \to P_3$ | $P_9 \to P_5$ | $P_9 \to P_6$ | $P_9 \to P_7$ | $P_9 \to P_8$ | $P_9 \to P_9$ | $P_9 \to P_{10}$ | $P_9 \to P_{11}$ | $P_9 \to P_{12}$ | $P_9 \to P_{13}$ | | |
| $P_{10}$ | $P_{10} \to P_5$ | $P_{10} \to P_6$ | $P_{10} \to P_7$ | $P_{10} \to P_8$ | $P_{10} \to P_9$ | $P_{10} \to P_{10}$ | $P_{10} \to P_{11}$ | $P_{10} \to P_{12}$ | $P_{10} \to P_{13}$ | $P_{10} \to P_{14}$ | | |
| $P_{11}$ | $P_{11} \to P_6$ | $P_{11} \to P_7$ | $P_{11} \to P_8$ | $P_{11} \to P_9$ | $P_{11} \to P_{10}$ | $P_{11} \to P_{11}$ | $P_{11} \to P_{12}$ | $P_{11} \to P_{15}$ | | | | |
| $P_{12}$ | $P_{12} \to P_7$ | $P_{12} \to P_8$ | $P_{12} \to P_9$ | $P_{12} \to P_{10}$ | $P_{12} \to P_{11}$ | $P_{12} \to P_{12}$ | $P_{12} \to P_{13}$ | $P_{12} \to P_{14}$ | $P_{12} \to P_{15}$ | $P_{12} \to P_{20}$ | | |
| $P_{13}$ | $P_{13} \to P_9$ | $P_{13} \to P_{10}$ | $P_{13} \to P_{11}$ | $P_{13} \to P_{12}$ | $P_{13} \to P_{13}$ | $P_{13} \to P_{14}$ | $P_{13} \to P_{15}$ | | | | | |
| $P_{14}$ | $P_{14} \to P_8$ | $P_{14} \to P_9$ | $P_{14} \to P_{11}$ | $P_{14} \to P_{12}$ | $P_{14} \to P_{13}$ | $P_{14} \to P_{14}$ | $P_{14} \to P_{15}$ | $P_{14} \to P_{16}$ | | | | |
| $P_{15}$ | $P_{15} \to P_{11}$ | $P_{15} \to P_{12}$ | $P_{15} \to P_{13}$ | $P_{15} \to P_{15}$ | $P_{15} \to P_{16}$ | $P_{15} \to P_{17}$ | | | | | | |
| $P_{16}$ | $P_{16} \to P_{13}$ | $P_{16} \to P_{14}$ | | | | | | | | | | |
| $P_{17}$ | $P_{17} \to P_{13}$ | $P_{17} \to P_{14}$ | | | | | | | | | | |
| $P_{18}$ | $P_{18} \to P_{17}$ | | | | | | | | | | | |
| $P_{19}$ | | | | | | | | | | | | |
| $P_{20}$ | $P_{20} \to P_{18}$ | | | | | | | | | | | |

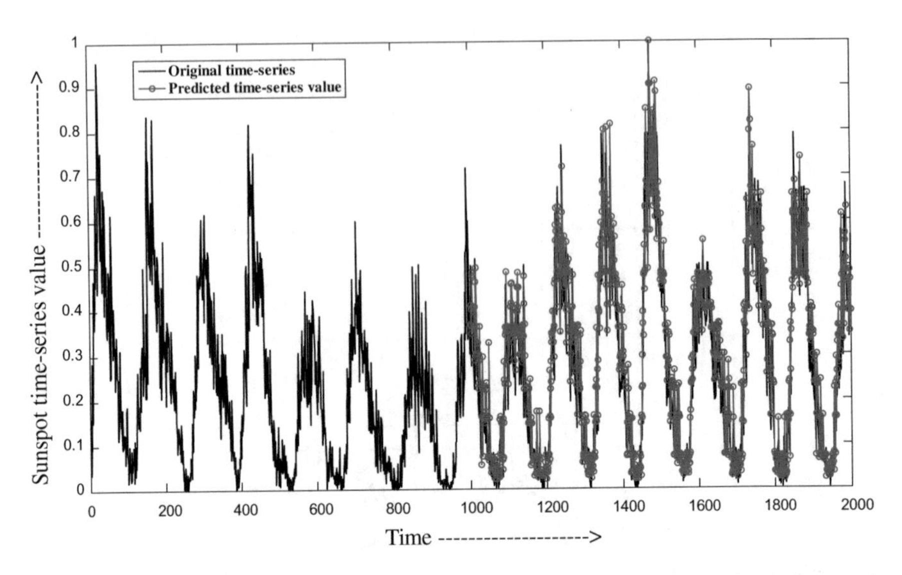


Fig. 5.7 Prediction of sunspot time-series using first-order transition rule based neural network model

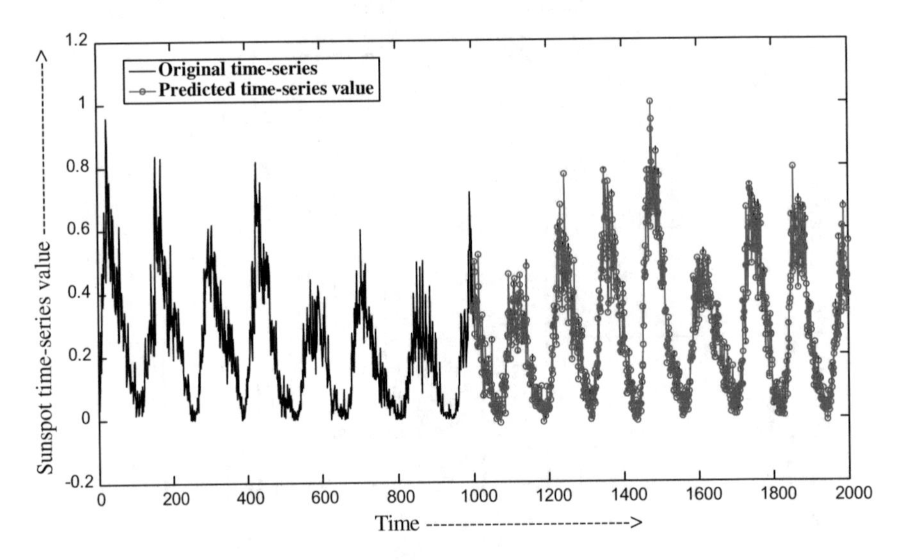


Fig. 5.8 Prediction of sunspot time-series using fuzzy rule based neural network model

$$NMSE = \frac{\sum_{t=1}^{N} \left(c_{test}(t) - c'(t)\right)^2}{\sum_{t=1}^{N} \left(c_{test}(t) - \bar{c}(t)\right)^2} \tag{5.20}$$

**Table 5.3** Comparison of prediction errors of existing methods in the literature with the proposed methods on the sunspot time-series

| Model | Prediction error | | |
|---|---|---|---|
| | MSE | RMSE | NMSE |
| Koskela et al. [39] | | | 9.79E−02 |
| Ma et al. [38] | | 1.29E−02 | 2.80E−03 |
| Smith et al. [17] | 2.32E−04 | 1.52E−02 | 9.46E−04 |
| Ardalani-Farsa et al. [18] | 1.4078E−4 | 0.0119 | 5.9041E−04 |
| **Proposed method 1** | **0.843E−04** | **0.0071** | **4.09E-04** |
| **Proposed method 2** | **0.094E−04** | **8.7472E−04** | **3.0315E-04** |

where $N$ is the total number of data points in the test phase time-series and $\bar{c}(t)$ is the average of all $N$ data points. In Table 5.3, we present a comparison in the prediction performance of the proposed approaches with respect to other existing neural models in the literature.

It is evident from Table 5.3 that the proposed models outperform other competitive models existing in the literature by a relatively large margin. Furthermore, among the two proposed methods, the second method which uses fuzzified first-order transition rules for training, leads to comparatively better prediction performance. This is primarily because of the fact that fuzzy representation of time-series data points lowers approximation errors in comparison to discrete quantization of data points with respect to respective partition mid-points. It can thus be concluded that the proposed models with their high prediction accuracy can be efficiently utilized for prediction of chaotic time-series.

### 5.5.2 *Experiment 2: TAIEX Close-Price Prediction*

In this experiment we apply the proposed models for prediction of the TAIEX close price time-series for the period 1990–2004. For each year, the time-series is divided into two periods: (i) the training period from January to October and (ii) the testing period from November to December. The following steps are carried out for the experiments:

*Step 1. Training Phase: Partitioning, First-order rule extraction and neural network training*. Following Sects. 5.3 and 5.4, the training period time-series is first partitioned. For partitioning, we have experimentally chosen 40 equi-spaced partitions as that yields the best prediction results. The first-order transition rules thus extracted from the time-series for each year is then segregated into training sets for the neural networks and modified into mid-point to mid-point mappings for the discrete rule based model and into membership values to mid-point mappings for

**Table 5.4** Comparison of proposed models with existing methods for the period 1990–1999

| Years models | 1990 | 1991 | 1992 | 1993 | 1994 | 1995 | 1996 | 1997 | 1998 | 1999 | Average |
|---|---|---|---|---|---|---|---|---|---|---|---|
| 1. Conventional models [23] | 220 | 80 | 60 | 110 | 112 | 79 | 54 | 148 | 167 | 149 | 117.9 |
| 2. Weighted models [23] | 227 | 61 | 67 | 105 | 135 | 70 | 54 | 133 | 151 | 142 | 114.5 |
| 3. Chen and Chen [27] | 172.89 | 72.87 | 43.44 | 103.21 | 78.63 | 66.66 | 59.75 | 139.68 | 124.44 | 115.47 | 97.70 |
| 4. Chen et al. [26] | 174.62 | 43.22 | 42.66 | 104.17 | 94.6 | 54.24 | 50.5 | 138.51 | 117.87 | 101.33 | 92.17 |
| 5. Chen and Kao [28] | **156.47** | 56.50 | 36.45 | 126.45 | 62.57 | 105.52 | 51.50 | 125.33 | **104.12** | 87.63 | 91.25 |
| 6. Cai et al. [29] | 187.10 | 39.58 | 39.37 | 101.80 | 76.32 | 56.05 | **49.45** | 123.98 | 118.41 | 102.34 | 89.44 |
| **7. Proposed method 1** | 163.13 | **33.88** | 40.63 | 103.94 | 78.43 | 62.75 | 49.62 | 126.21 | 106.34 | 91.27 | 85.62 |
| **8. Proposed method 2** | 158.26 | 36.41 | **34.38** | **98.73** | **61.14** | **50.48** | 51.16 | **121.67** | 112.44 | **87.03** | **81.17** |

**Table 5.5** Comparison of proposed models with existing methods for the period 1999–2004

| Years methods | 1999 | 2000 | 2001 | 2002 | 2003 | 2004 | Average |
|---|---|---|---|---|---|---|---|
| 1. Huarng et al. [25]. (Using NASDAQ) | NA | 158.7 | 136.49 | 95.15 | 65.51 | 73.57 | 105.88 |
| 2. Huarng et al. [25]. (Using Dow Jones) | NA | 165.8 | 138.25 | 93.73 | 72.95 | 73.49 | 108.84 |
| 3. Huarng et al. [25]. (Using $M_{1b}$) | NA | 160.19 | 133.26 | 97.1 | 75.23 | 82.01 | 111.36 |
| 4. Huarng et al. [25]. (Using NASDAQ and $M_{1b}$) | NA | 157.64 | 131.98 | 93.48 | 65.51 | 73.49 | 104.42 |
| 5. Huarng et al. [25]. (Using Dow Jones and $M_{1b}$) | NA | 155.51 | 128.44 | 97.15 | 70.76 | 73.48 | 105.07 |
| 6. Huarng et al. [25]. (Using NASDAQ,Dow Jones and $M_{1b}$) | NA | 154.42 | 124.02 | 95.73 | 70.76 | 72.35 | 103.46 |
| 7. Chen et al. [30, 31, 32] | 120 | 176 | 148 | 101 | 74 | 84 | 117.4 |
| 8.U_R Model [31, 32] | 164 | 420 | 1070 | 116 | 329 | 146 | 374.2 |
| 9.U_NN Model [31, 32] | 107 | 309 | 259 | 78 | 57 | 60 | 145.0 |
| 10.U_NN_FTS Model [31, 32, 33] | 109 | 255 | 130 | 84 | 56 | 116 | 125.0 |
| 11.U_NN_FTS_S Model [31, 32, 33] | 109 | 152 | 130 | 84 | 56 | 116 | 107.8 |
| 12.B_R Model [31, 32] | 103 | 154 | 120 | 77 | 54 | 85 | 98.8 |
| 13.B_NN Model [31, 32] | 112 | 274 | 131 | 69 | 52 | 61 | 116.4 |
| 14.B_NN_FTS Model [31, 32] | 108 | 259 | 133 | 85 | 58 | 67 | 118.3 |
| 15.B_NN_FTS_S Model [31, 32] | 112 | 131 | 130 | 80 | 58 | 67 | 96.4 |
| 16. Chen and Chen [27]. (Using Dow Jones) | 115.47 | 127.51 | 121.98 | 74.65 | 66.02 | 58.89 | 94.09 |
| 17. Chen and Chen [27]. (Using NASDAQ) | 119.32 | 129.87 | 123.12 | 71.01 | 65.14 | 61.94 | 95.07 |
| 18. Chen and Chen [27]. (Using $M_{1b}$) | 120.01 | 129.87 | 117.61 | 85.85 | 63.1 | 67.29 | 97.29 |
| 19. Chen and Chen [27]. (Using NASDAQ and Dow Jones) | 116.64 | 123.62 | 123.85 | 71.98 | 58.06 | 57.73 | 91.98 |
| 20. Chen and Chen [27]. (Using Dow Jones and $M_{1b}$) | 116.59 | 127.71 | 115.33 | 77.96 | 60.32 | 65.86 | 93.96 |
| 21. Chen and Chen [27]. (Using NASDAQ and $M_{1b}$) | 114.87 | 128.37 | 123.15 | 74.05 | 67.83 | 65.09 | 95.56 |
| 22. Chen and Chen [27]. (Using NASDAQ, Dow Jones and $M_{1b}$) | 112.47 | 131.04 | 117.86 | 77.38 | 60.65 | 65.09 | 94.08 |
| 23. Karnik-Mendel [34] induced stock prediction | 116.60 | 128.46 | 120.62 | 78.60 | 66.80 | 68.48 | 96.59 |
| 24. Chen et al. [26]. (Using NASDAQ, Dow Jones and $M_{1b}$) | 101.47 | 122.88 | 114.47 | 67.17 | 52.49 | 52.84 | 85.22 |
| 25. Chen and Kao [28] | 87.67 | 125.34 | 114.57 | 76.86 | 54.29 | 58.17 | 86.14 |
| 26. Cai et al. [29] | 102.22 | 131.53 | 112.59 | 60.33 | 51.54 | **50.33** | 84.75 |
| **27. Proposed method 1** | 91.27 | 120.16 | 108.63 | **59.18** | 46.88 | 91.62 | 86.29 |
| **28. Proposed method 2** | **87.03** | **102.04** | **99.49** | 61.25 | **39.14** | 70.11 | **76.51** |

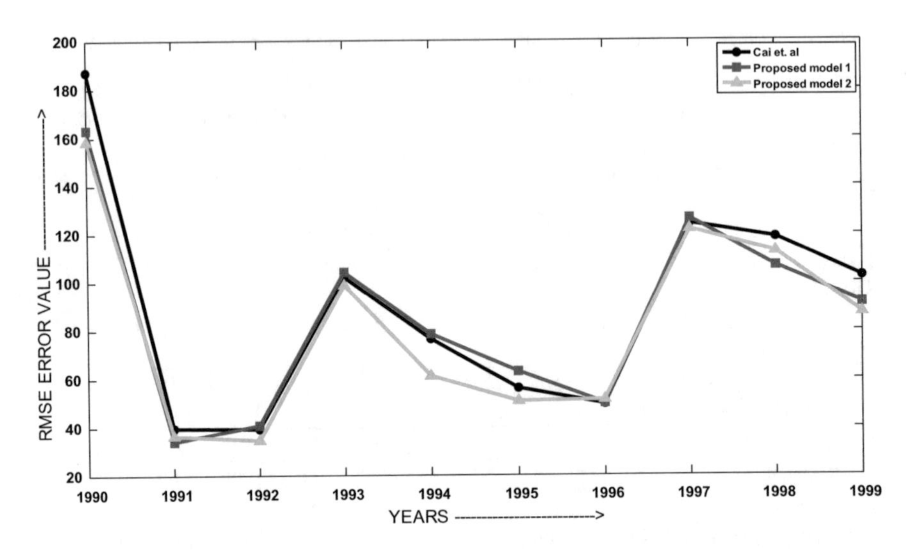


**Fig. 5.9** RMSE Error values for Cai et al. [29] and the two proposed models for the years 1990–1999

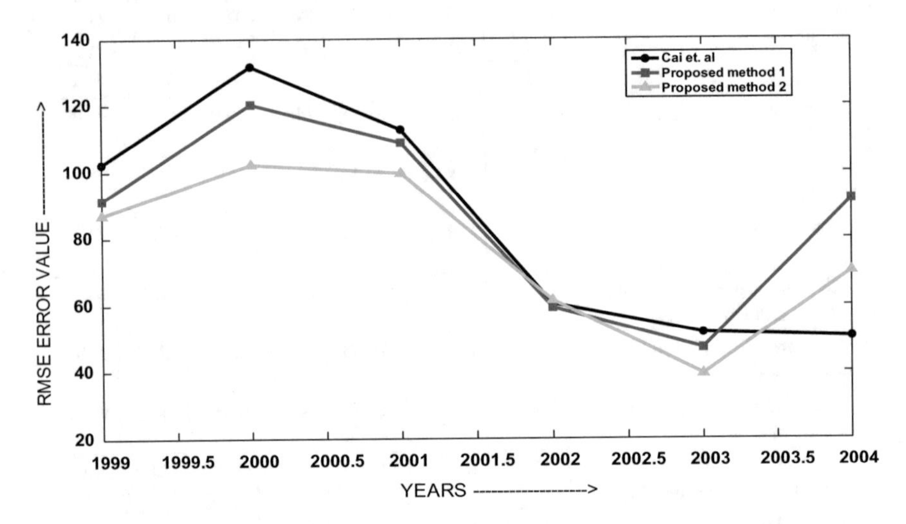


**Fig. 5.10** RMSE Error values for Cai et al. [29] and the two proposed models for the years 1990–2004

the fuzzy rule based model. The neural networks are trained using the back propagation algorithm on the training sets obtained above.

*Step 2. Testing Phase: Prediction on test series.* In the test phase, the trained neural networks of both the proposed models are used to make predictions on the time-series of the testing period for each year (1990–2004). In order to measure the prediction error of the models, we use the RMSE error metric as defined in Eq. (5.19). A comparative study of the proposed models with various existing models in the literature has been performed. For the period, 1990–1999, the proposed models are compared with conventional models [23], weighted models [23], Chen and Chen's model [27], Chen et. al's model [26], Chen and Kao's model [28] and Cai et al's model [29]. The results of all the above mentioned models are obtained from [29] and are shown in Table 5.4. For the period 1999–2004, the proposed models are compared with methods specified in [25–34] and the results are summarized in Table 5.5. The minimum RMSE values for each year have been shown in bold. Figs. 5.9 and 5.10 illustrate the variations in the RMSE values graphically over the years 1990–1999 and 1999–2004 respectively, for the proposed models along with the next best model [29].

As is evident from Tables 5.4 and 5.5, the proposed models clearly outperform the other existing models in the literature. The fuzzy rule-based neural network model outperforms the next best existing model by 9.25% for the period 1990–1999 and by 9.72% for the period 1999–2004. The results indicate that the proposed models can be effectively utilized in financial prediction.

## 5.6 Conclusion

In this chapter, we presented a novel grouping scheme of first-order transition rules obtained from a partitioned time-series for the purpose of fuzzy-induced neural network based prediction. In this direction, we have proposed two models. The first model uses first-order transition rules segregated into groups representing injective mappings from antecedents to consequents of a rule. Each rule in a group thus obtained possesses a distinct antecedent and each such group is used to train a separate neural network in the ensemble. This helps in realizing the simultaneous and concurrent firing of multiple rules during prediction. Furthermore, the individual predictions of the networks are weighted according to the probability of occurrence of their corresponding transition rules and the weighted sum thus obtained is treated as the final predicted value of the model. This helps in taking the recurrence of transition rules into account while making forecasts, thereby increasing prediction accuracy.

The second model proposed modifies and extends the training scheme of the first, by considering each partition of the time-series as a fuzzy set to identify the membership of a data point in its respective partition. The advantage of such an approach lies in utilizing the inherent fuzziness involved in identifying the partition to which a time-series data point belongs, thereby reducing the approximation error induced due to quantization of a time-series data point with respect to its partition mid-point value. The first-order transition rules are thus converted into fuzzy first-order rules and segregated into training sets following the grouping scheme as mentioned above.

Extensive experiments carried out on real life chaotic time-series like the sunspot [36] as well as on economic time-series like the TAIEX [37] reveal a high prediction accuracy for both the proposed models. Further, it is also observed that the performance of the fuzzy rule based neural network ensemble is comparatively better than its predecessor in both the experiments carried out. Prediction performance can possibly be further increased by approaches like optimized partitioning, higher order transition rules for training and chronological weight assignment of transition rules. Such approaches form a future scope for work in this regard. Thus, with the high forecast accuracy and low prediction error of the proposed models compared to other existing models in the literature, we can conclude that the said models can be effectively utilized for real-life time-series forecasting applications.

## Exercises

1. Let a time-series be partitioned into seven disjoint partitions: $P_1$, $P_2$, $P_3$, … $P_7$. The rules acquired from the time-series include $S_1 \cup S_2 \cup S_3$, where,

$$S_1 = \{P_3 \rightarrow P_5, P_4 \rightarrow P_3, P_5 \rightarrow P_2\}$$
$$S_2 = \{P_3 \rightarrow P_4, P_4 \rightarrow P_6, P_5 \rightarrow P_4\}$$
$$S_3 = \{P_4 \rightarrow P_7, P_5 \rightarrow P_1, P_2 \rightarrow P_3\}$$

we realize three sets: $S_1$, $S_2$, and $S_3$ disjointly on neural nets. Suppose for input being $P_4$, the prediction rule is obtained as

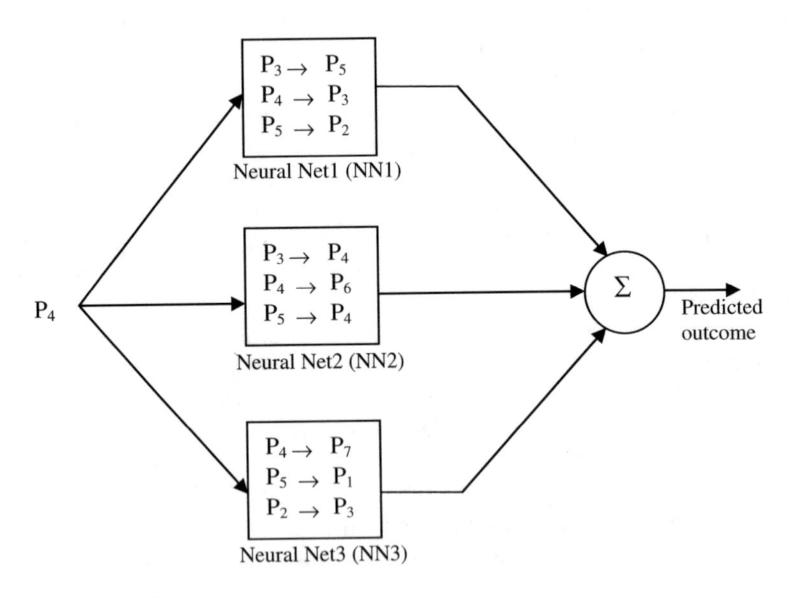


**Fig. 5.11** Problem 1

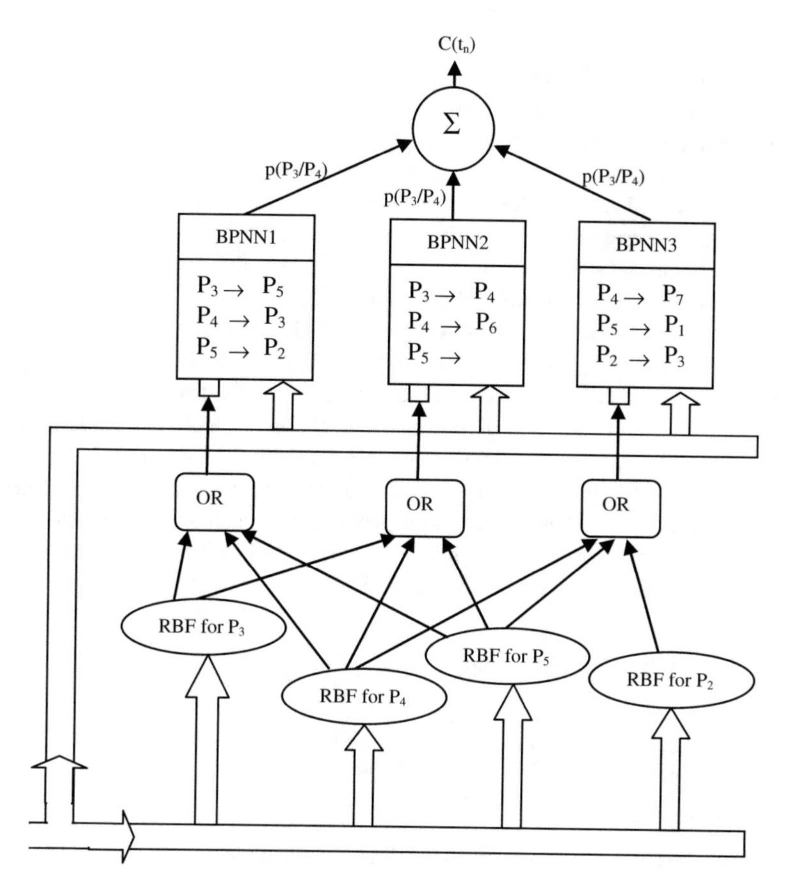


**Fig. 5.12** Problem 3

$$\begin{aligned}
&\text{Expected Value of Predicted Stock} - \text{Price} \\
&\quad = \text{Mid_value_of_Partition_P}_3 \times \text{Prob}(P_3/P_4) \\
&\quad + \text{Mid_value_of_Partition_P}_6 \times \text{Prob}(P_6/P_4) \\
&\quad + \text{Mid_value_of_Partition_P}_7 \times \text{Prob}(P_7/P_4)
\end{aligned}$$

Let the mid values of partitions $P_3$, $P_6$, and $P_7$ are 1K, 2K and 3K respectively. Let $\text{Prob}(P_3/P_4) = \text{Prob}(P_6/P_4) = \text{Prob}(P_7/P_4) = 0.33$, Evaluate expected value of predicted stock-price.

2. In Q. 1, if today's price falls in partition $P_4$, what efficient relation would you suggest for time-efficient prediction? Fig. 5.11

[**Hints:** Since $P_4$ occurs in the left of the rules: $P_4 \rightarrow P_3$, $P_4 \rightarrow P_6$, $P_4 \rightarrow P_7$, to efficiently predict the next partition, we need to realize the last three rule on 3 different neural nets. Thus the three neural nets will be triggered with input = $P_4$, to produce their resulting response].

3. Suppose, we have three sets of disjoint rules: $S_1$, $S_2$, and $S_3$ as given in Q2. These set of rules are realized on three feed forward neural nets. We would like to trigger a neural net, if the partition appearing at the left of →operator falls in a rule in the neural net. The occurrence of the left symbol (partition) can be tested using Radial Basis Function (RBF) neural net.
   Design the architecture of a system, where RBF neurons select a feed forward structure depending on the occurrence of a partition in the left side of → operator. The predicted price may be evaluated by the formulation in problem.
   [**Hints:** The architecture is given below. In the given architecture, RBF neuron select the appropriate neural nets containing the supplied partition at the left side of → operator. The selected neural nets are fired to produce the right hand side of the rules referred to above. The product of the mid-value of the inferred partition for the proposed neural nets is collected and the expectation of the targeted close price is evaluated.] Fig. 5.12

## Appendix 5.1: Source Codes of the Programs

% MATLAB Source Code of the Main Program and Other Functions for Time-%Series Prediction by Fuzzy-induced Neural Regression

% Developed by Jishnu Mukhoti

% Under the guidance of Amit Konar and Diptendu Bhattacharya

```
%% Main 0
function [ partitions ] = main_0( time_series, num_part )
%UNTITLED6 Summary of this function goes here
%  Detailed explanation goes here

partitions = partition(time_series,num_part);

end
%%%%%%%%%%%%%%%%%%%%
%% Main 1

function [ rules ] = main_1( time_series, partitions )
%Given a time-series and number of partitions provides the transition rules
using sub
%functions.
```

```
%plot_partitions(time_series,partitions);
rules = find_transition_rules(time_series, partitions);

end
%%%%%%%%%%%%%%%%%
%% Main 2

function [ refined_training_set1, refined_train-
ing_set2, rule_prob ] = main_2( rules, partitions )
%A main function to prepare the training set and refine them for the neural
%net training.

training_set = create_training_set(rules)
refined_training_set1 = refine_training_set_part_1(training_set,parti-
tions);
refined_training_set2 = refine_training_set_part_2(refined_training_set1,-
partitions);
rule_prob = rule_probability(rules);
end
%%%%%%%%%%%%%%%%%
%% Create training set

function [ res ] = create_training_set( rules )
%From the extracted rules of the time-series, this function produces the
%training set to train the neural networks.

r2 = rules;
num_part = size(rules,1);

%Count of neural networks required to train based on the given rules
nn_count = 1;
flag = 0;

while flag == 0
  index = 1;
  flag = 1;
  for i = 1:num_part
    for j = 1:num_part
      if (r2(i,j) ~= 0)
        res(index,1,nn_count) = i;
        res(index,2,nn_count) = j;
        r2(i,j) = 0;
        index = index + 1;
        flag = 0;
```

```
            break;
         end
      end
   end
   nn_count = nn_count + 1;
end
end
%%%%%%%%%%%%%%%%%%%%%%%%%%
%% Create training set part 2

function [ ts ] = create_training_set_part2( train_series, partitions )
%UNTITLED2 Summary of this function goes here
%  Detailed explanation goes here

l = length(train_series);
num_part = size(partitions,1);

ts = zeros(l-1, num_part+1);

for i = 1:l-1
   mv = gauss_mf(partitions,train_series(i));
   ts(i,1:end-1) = mv;
   ts(i,end) = train_series(i+1);
end

end
%%%%%%%%%%%%%%%%%%%%%%%%
%% Error Matrices

function [ rms, ms, nms ] = error_metrics( actual, pred )
%Function to compute the MSE, RMSE and NMSE errors.

rms = rmse(actual, pred);
ms = rms^2;
nms = nmse(pred, actual);

end
%%%%%%%%%%%%%%%%%%%%%
%% Find partitions

function [ res ] = find_transition_rules( series, partitions )
% Finds the transition rules given the time-series and its partitions
```

```
num_part = size(partitions,1);
len = length(series);

res = zeros(num_part,num_part);

for i = 1:len-1
  prev_part = part_data_pt(partitions,series(i));
  next_part = part_data_pt(partitions,series(i+1));
  res(prev_part,next_part) = res(prev_part,next_part) + 1;
end
end
%%%%%%%%%%%%%%%%%%%%%%%%%%%%%
%% Gaussian Membership Functions

function [ mem_val ] = gauss_mf( partitions, point )
%A function to take a point and return the membership values in all the
%membership functions.

num_part = size(partitions,1);
mem_val = zeros(num_part,1);

for i = 1:num_part
  mean = (partitions(i,1) + partitions(i,2))/2;
  sig = partitions(i,2) - mean;
  mem_val(i) = gaussmf(point, [sig mean]);
end
end
%%%%%%%%%%%%%%%%%%%%%%%%%
% lorenz - Program to compute the trajectories of the Lorenz

% equations using the adaptive Runge-Kutta method.
clear;  help lorenz;

%* Set initial state x,y,z and parameters r,sigma,b
state = input('Enter the initial position [x y z]: ');
r = input('Enter the parameter r: ');
sigma = 10.;  % Parameter sigma
b = 8./3.;    % Parameter b
param = [r sigma b]; % Vector of parameters passed to rka
tau = 1;      % Initial guess for the timestep
err = 1.e-3;  % Error tolerance

%* Loop over the desired number of steps
time = 0;
```

```
nstep = input('Enter number of steps: ');
for istep=1:nstep

 %* Record values for plotting
 x = state(1); y = state(2); z = state(3);
 tplot(istep) = time;  tauplot(istep) = tau;
 xplot(istep) = x;  yplot(istep) = y;  zplot(istep) = z;
 if( rem(istep,50) < 1 )
  fprintf('Finished %g steps out of %g\n',istep,nstep);
 end

 %* Find new state using adaptive Runge-Kutta
 [state, time, tau] = rka(state,time,tau,err,'lorzrk',param);

end

%* Print max and min time step returned by rka
fprintf('Adaptive time step: Max = %g,  Min = %g \n', ...
       max(tauplot(2:nstep)), min(tauplot(2:nstep)));

%* Graph the time series x(t)
figure(1); clf;  % Clear figure 1 window and bring forward
plot(tplot,xplot,'-')
xlabel('Time');  ylabel('x(t)')
title('Lorenz model time series')
pause(1)  % Pause 1 second

%* Graph the x,y,z phase space trajectory
figure(2); clf;  % Clear figure 2 window and bring forward
% Mark the location of the three steady states
x_ss(1) = 0;            y_ss(1) = 0;       z_ss(1) = 0;
x_ss(2) = sqrt(b*(r-1));  y_ss(2) = x_ss(2); z_ss(2) = r-1;
x_ss(3) = -sqrt(b*(r-1)); y_ss(3) = x_ss(3); z_ss(3) = r-1;
plot3(xplot,yplot,zplot,'-',x_ss,y_ss,z_ss,'*')
view([30 20]);  % Rotate to get a better view
grid;          % Add a grid to aid perspective
xlabel('x'); ylabel('y'); zlabel('z');
title('Lorenz model phase space');
%%%%%%%%%%%%%%%%%%%%%%%%%%
%% NMSE Calculation

 function [ err ] = nmse( vec1, vec2 )
 %Function to compute the normalized mean square error.
```

```
v = abs(vec1 - vec2);
v = v.^2;
s1 = sum(v);
s2 = sum(abs(vec1 - mean(vec2)));
err = s1/s2;

end
%%%%%%%%%%%%%%%%%%%%%%%%
%% Script to create, execute and test the neural net model.

%
% Creating the data sets and partitioning it into sets of 1000 data points %
%
clear; close all; clc;
load 'sunspot.txt';
data = sunspot(:,4);
run = ceil(size(data)/1000);
data_sets = zeros(length(data),run);
j = 1;
% for i = 1:run
%    data_sets(:,i) = data(j:j+999);
%    j = j + 1000;
% end
data_sets(:,1) = data;

%% Training neural net and predicting for each test run of the data %%

num_part = 20;
rmse_val = zeros(run,1);
%TODO: Convert the second 1 to run !!!!
for i = 1:1
   %Separate the series into training and testing periods
   dt = data_sets(:,i);
   l = floor(0.5*length(dt));
   train_series = dt(1:l);
   test_series = dt(l+1:end);
   partitions = main_0(dt,num_part);
   plot_partitions(train_series, partitions);
   rules = main_1(train_series, partitions);
   [rts1, rts2, rule_prob] = main_2(rules, partitions);
   fprintf('Training the neural networks for part 1\n');
   nets1 = train_neural_nets(rts1);
  % fprintf('Training the neural networks for part 2\n');
```

```
% nets2 = train_neural_nets2(rts2);

 %Prediction phase
 pred11 = zeros(length(dt)-1,1);
 pred12 = zeros(length(dt)-1,1);
 %pred2 = zeros(200,1);
 fprintf('Running test cases ..................\n');
 for j = 1:(length(dt)-1)
    fprintf('———Iteration %d——————————\n',j-1+1);
    inp = dt(j);
    [out11,out12] = prediction(inp,rule_prob,nets1,partitions);
    %out2 = prediction2(inp,rule_prob,nets2,partitions);
    pred11(j-1+1) = out11;
    pred12(j-1+1) = out12;
    %pred2(j-799) = out2;
  end
  rmse_val(i) = rmse(test_series, pred11);
  %Plot the predictions
   figure;
   plot((1:(length(dt)-1))',test_series,'k*-');
   hold on;
   plot((1:(length(dt)-1))',pred11,'r*-');
  %plot((1:100)',pred12,'b*-');
  %plot((1:200)',pred2,'b*-');
  end
%%%%%%%%%%%%%%%%%%%%%%
%% Loading data and preparing the training set %%

close all; clear; clc;

load 'data.txt';
run = ceil(size(data)/1000);
data_sets = zeros(length(data),run);
j = 1;
% for i = 1:run
%    data_sets(:,i) = data(j:j+999);
%    j = j + 1000;
% end
data_sets(:,1) = data;
%% Train the neural network %%

num_part = 40;
rmse_val = zeros(run,1);
```

```
for i = 1:run
   %Separate the series into training and testing periods
   dt = data_sets(:,i);
   l = floor(0.8*length(dt));
   train_series = dt(1:l);
   test_series = dt(l+1:end);
   partitions = main_0(dt,num_part);

   ts = create_training_set_part2(train_series, partitions);
   net = train_part2(ts);

   %Prediction phase
   preds = zeros(length(dt)-l,1);

   for j = l:(length(dt)-1)
      fprintf('———Iteration %d————————\n',j-l+1);
      inp = dt(j);
      preds(j-l+1) = predict_part2(net,inp,partitions);
   end

   %Calculate rmse and plot%
   rmse_val(i) = rmse(test_series, preds);
   figure;
   plot((1:(length(dt)-l))',test_series,'k*-');
   hold on;
   plot((1:(length(dt)-l))',preds,'r*-');
end
%%%%%%%%%%%%%%%%%%
%% Part data partition

function [ res ] = part_data_pt( partitions, point )
%A function to find the partition to which a data point belongs.

res = 0;
num_part = size(partitions, 1);

for i = 1:num_part
   if ((point >= partitions(i,1)) && (point <= partitions(i,2)))
      res = i;
      break;
   end
end
```

```
end
%%%%%%%%%%%%%%%%%%%%
%% Partitioning

function [ res ] = partition( series, num_part )
%A function to partition the given time series into the number of
%partitions specified as a parameter

mx = max(series);
mn = min(series);

diff = mx-mn;
part_width = diff/num_part;

res = zeros(num_part,2);

temp = mn;

for i = 1:num_part
   res(i,1) = temp;
   temp = temp + part_width;
   res(i,2) = temp;
end
end
%%%%%%%%%%%%%%%%
%% Plotting partitions

function [  ] = plot_partitions( series, partitions )
%Plots the time-series and its partitions

plot([1:length(series)]',series,'k*-');
hold on;

for i = 1:(size(partitions,1))
   line([1,length(series)],[partitions(i,1),partitions(i,1)]);
end

n = size(partitions,1);
line([1,length(series)],[partitions(n,2),partitions(n,2)]);

end
%%%%%%%%%%%%%
```

```
%
% Given a time-series data point, the function uses the trained neural nets
%% to make a future prediction.

function [ s1,s2 ] = prediction( point, rule_prob, nets, partitions )

nn_count = size(nets,2);
prev_part = part_data_pt(partitions,point);
num_part = size(partitions,1);
s = 0;

preds = zeros(nn_count,1);
preds2 = zeros(nn_count,1);
probs = zeros(nn_count,1);

% fprintf ('————————————————————————————————————————\n');
% fprintf('Input given: %f\n',point);
% fprintf('Input partition: %d\n',prev_part);
for i = 1:nn_count
  %fprintf ('Output for neural network %d\n',i);
  pred = nets(i).net(point);
  preds2(i) = pred;
  if (pred > partitions(num_part,2))
    pred = partitions(num_part,2);
  end
  if (pred < partitions(1,1))
    pred = partitions(num_part,1);
  end
  %fprintf('Prediction: %f\n',pred);
  next_part = part_data_pt(partitions,pred);
  %fprintf('Output partition: %d\n',next_part);
  prob = rule_prob(prev_part, next_part);
  %fprintf('Probability of transition: %f\n',prob);
  %s = s + (pred * prob);
  %fprintf('current value of overall prediction: %f\n', s);
  preds(i) = pred;
  probs(i) = prob;
end

%Process the prob vector%
mx = sum(probs);
if mx ~= 0
  probs = probs/mx;
else
```

```
  for i = 1:nn_count
    probs(i) = 1/nn_count;
  end
end

% for i = 1:nn_count
%    fprintf ('Output for neural network %d\n',i);
%    fprintf('Prediction: %f\n',preds(i));
%    fprintf('Probability of transition: %f\n',probs(i));
% end

s1 = preds .* probs;
s1 = sum(s1);
s2 = mean(preds2);
% fprintf ('Value of overall prediction by weightage : %f\n', s1);
% fprintf('Value of overall prediction by simple average: %f\n', s2);
%pause(1);
end

%%%%%%%%%%%%%%%%%%%%%%%%%%
%
% Given a time-series data point, the function uses the trained neural nets
%% to make a future prediction.

function [ s1, s2 ] = prediction2( point, rule_prob, nets, partitions )
%Given a time-series data point, the function uses the trained neural nets
%to make a future prediction.

nn_count = size(nets,2);
prev_part = part_data_pt(partitions,point);
num_part = size(partitions,1);

preds = zeros(nn_count,1);
preds2 = zeros(nn_count,1);
probs = zeros(nn_count,1);

fprintf ('————————————————————————————————————————————————————————\n');
fprintf('Input given: %f\n',point);
fprintf('Input partition: %d\n',prev_part);
for i = 1:nn_count
   fprintf ('Output for neural network %d\n',i);
   mv = gauss_mf(partitions,point);
   pred = nets(i).net(mv);
   preds2(i) = pred;
```

```
    if (pred > partitions(num_part,2))
        pred = partitions(num_part,2);
    end
    if (pred < partitions(1,1))
        pred = partitions(num_part,1);
    end
    %fprintf ('Prediction by neural net 2 : %lf\n',pred);
    next_part = part_data_pt(partitions,pred);
    prob = rule_prob(prev_part, next_part);
    preds(i) = pred;
    probs(i) = prob;
end

%Process the prob vector%
mx = sum(probs);
if mx ~= 0
    probs = probs/mx;
else
    for i = 1:nn_count
        probs(i) = 1/nn_count;
    end
end

for i = 1:nn_count
    fprintf ('Output for neural network %d\n',i);
    fprintf('Prediction: %f\n',preds(i));
    fprintf('Probability of transition: %f\n',probs(i));
end

s1 = preds .* probs;
s1 = sum(s1);
s2 = mean(preds2);

fprintf ('Value of overall prediction by weightage : %f\n', s1);
fprintf('Value of overall prediction by simple average: %f\n', s2);
%pause(1);

end

%%%%%%%%%%%%%%%%%%%%
%% Prediction part2

function [ res ] = predict_part2( net, point, partitions )
%UNTITLED4 Summary of this function goes here
```

```
% Detailed explanation goes here

mv = gauss_mf(partitions, point);
res = net(mv);
end
%%%%%%%%%%%%%%%%%%
% Function refines the training set to form mid value to mid value mapping.

function [ res ] = refine_training_set_part_1( training_set, partitions )

num_part = size(partitions,1);
mid_vals = zeros(num_part, 1);

for i = 1:num_part
  mid_vals(i) = (partitions(i,1) + partitions(i,2))/2;
end

nn_count = size(training_set,3);
rows = size(training_set,1);
res = zeros(size(training_set));

for i = 1:nn_count
  train = training_set(:,:,i);
  for j = 1:rows
    if train(j,1) ~= 0
      prev = train(j,1);
      next = train(j,2);
      train(j,1) = mid_vals(prev);
      train(j,2) = mid_vals(next);
    end
  end
  res(:,:,i) = train;
end
end

%%%%%%%%%%%%%%%%%%%%%%%
%A function to produce a training set for training a neural net using fuzzy
%membership values of the input time-series value.

function [ res ] = refine_training_set_part_2( ref_training_set, parti-
tions )

nn_count = size(ref_training_set,3);
rows = size(ref_training_set,1);
```

```
num_part = size(partitions,1);
res = zeros(rows,num_part+1,nn_count);

for i = 1:nn_count
   tr = ref_training_set(:,:,i);
   for j = 1:rows
      if tr(j,1) ~= 0
         res(j,1:end-1,i) = (gauss_mf(partitions,tr(j,1)))';
         res(j,end,i) = tr(j,2);
      end
   end
end
end

%%%%%%%%%%%%%%%%%%%%%%
%Find the rmse of vec1 and vec2

function [ res ] = rmse( vec1, vec2 )

v = abs(vec1 - vec2);
v = v.^2;
m = mean(v);
res = sqrt(m);

end
%%%%%%%%%%%%%%%%%%%%%%%%
%A function to convert the rule matrix to a transition probability matrix.

function [ res ] = rule_probability( rules )

num_part = size(rules,1);
s = sum(rules,2);
res = zeros(size(rules));

for i = 1:num_part
   if s(i) ~= 0
      res(i,:) = rules(i,:)/s(i);
   else
      res(i,:) = 0;
   end
end
end
```

```
%%%%%%%%%%%%%%%%%%%%
%A function to train the neural networks on the given data.

function [ a ] = train_neural_nets( refined_training_set )

nn_count = size(refined_training_set,3);
r = size(refined_training_set,1);

nn_rc = 0;
for i = 1:nn_count
   tr = refined_training_set(:,:,i);
   idx = 1;
   while idx <= r
      if tr(idx,1) == 0
         break;
      end
      idx = idx + 1;
   end
   if idx >= 5
      nn_rc = nn_rc + 1;
   end
end
nn_count = nn_rc;

for i = 1:nn_count
   %Prepare the training data
   tr = refined_training_set(:,:,i);
   idx = 1;
   while idx <= r
      if tr(idx,1) == 0
         break;
      end
      idx = idx +1;
   end
   tr = tr(1:idx-1,:);
   %Code for neural net
   a(i).net = feedforwardnet(10);
   a(i).net = train(a(i).net,(tr(:,1))',(tr(:,2))');
end
end

%%%%%%%%%%%%%%%%%%%
%A function to train the neural networks on the given data.
```

```
function [ a ] = train_neural_nets2( refined_training_set )

nn_count = size(refined_training_set,3);
r = size(refined_training_set,1);
num_part = size(refined_training_set,2) - 1;

nn_rc = 0;
for i = 1:nn_count
  tr = refined_training_set(:,:,i);
  idx = 1;
  while idx <= r
    s = sum(tr(idx,1:end-1));
    if s == 0
      break;
    end
    idx = idx + 1;
  end
  if idx >= 5
    nn_rc = nn_rc + 1;
  end
end
nn_count = nn_rc;

for i = 1:nn_count
  %Prepare the training data
  tr = refined_training_set(:,:,i);
  idx = 1;
  while idx <= r
    s = sum(tr(idx,1:end-1));
    if s == 0
      break;
    end
    idx = idx +1;
  end
  tr = tr(1:idx-1,:);
  %Code for neural net
  a(i).net = feedforwardnet(num_part+10);
  a(i).net = train(a(i).net,(tr(:,1:end-1))',(tr(:,end))');
end
end
%%%%%%%%%%%%%%%%
%% Training Part 2 for training data set
```

```
function [ net ] = train_part2( ts )
%UNTITLED3 Summary of this function goes here
%  Detailed explanation goes here
num_part = size(ts,2) - 1;
net = feedforwardnet(num_part + 10);
net = train(net, (ts(:,1:end-1))',(ts(:,end))');
end
%%%%%%%%%%%%%%%%%%%%%%
```

## References


1. Chen, S. M., & Hwang, J. R. (2000). Temperature prediction using fuzzy time series. *IEEE Transactions on Systems, Man, Cybernetics, Part-B, 30*(2), 263–275.
2. Wu, C. L., & Chau, K. W. (2013). Prediction of rainfall time series using modular soft computing methods. *Elsevier, Engineering Applications of Artificial Intelligence, 26,* 997–1007.
3. Morales-Esteban, A., Martinez-Alvarez, F., Troncoso, A., Justo, J. L., & Rubio-Escudero, C. (2010). Pattern recognition to forecast seismic time series. *Elsevier, Expert Systems with Applications, 37,* 8333–8342.
4. Barnea, O., Solow, A. R., & Stone, L. (2006). On fitting a model to a population time series with missing values. *Israel Journal of Ecology and Evolution, 52,* 1–10.
5. Jalil, A., & Idrees, M. (2013). Modeling the impact of education on the economic growth: evidence from aggregated and disaggregated time series data of Pakistan. *Elsevier Economic Model, 31,* 383–388.
6. Box, G. E. P., & Jenkins, G. (1976). *Time series analysis, forecasting and control.* San Francisco, CA: Holden-Day.
7. Wang, C. C. (2011). A comparison study between fuzzy time series model and ARIMA model for forecasting Taiwan Export. *Elsevier, Expert Systems with Applications, 38*(8), 9296–9304.
8. Chang, B. R., & Tsai, H. F. (2008). Forecast approach using neural network adaptation to support vector regression grey model and generalized auto-regressive conditional heteroscedasticity. *Elsevier, Expert Systems with Applications, 34*(2), 925–934.
9. Tsokos, C. P. (2010). K-th moving, weighted and exponential moving average for time series forecasting models. *European Journal of Pure and Applied Mathematics, 3*(3), 406–416.
10. Zhang, G. P. (2003). Time series forecasting using a hybrid ARIMA and neural network model. *Elsevier, Neurocomputing, 50,* 159–175.
11. Zadeh, L. A. (1965). Fuzzy sets. *Information and Control, 8,* 338–353.
12. Zhang, G. P. (2000). Neural networks for classification: A survey. *IEEE Transactions on Systems, Man, Cybernetics Part C (Applications and Reviews), 30*(4), 451–462.
13. Ma, L., & Khorasani, K. (2004). New training strategies for constructive neural networks with application to regression problems. *Elsevier, Neural Networks, 17*(4), 589–609.
14. Hill, M., O' Connor, T., & Remus, W. (1996). Neural network models for time series forecasts. *Management Science, 42*(7), 1082–1092.
15. Hamzacebi, C. (2008). Improving artificial neural networks' performance in seasonal time series forecasting. *Elsevier, Information Sciences, 178,* 4550–4559.
16. Mirikitani, D. T., & Nikolaev, N. (2010). Recursive Bayesian recurrent neural networks for time-series modeling. *IEEE Transactions on Neural Networks, 21*(2), 262–274.
17. Smith, C., & Jin, Y. (2014). Evolutionary multi-objective generation of recurrent neural network ensembles for time series prediction. *Elsevier, Neurocomputing, 143,* 302–311.

18. Ardalani-Farsa, M., & Zolfaghari, S. (2010). Chaotic Time series prediction with residual analysis method using hybrid Elman-NARX neural networks. *Elsevier, Neurocomputing, 73,* 2540–2553.
19. Gaxiola, P., Melin, F., & Valdez, F. (2014). Interval type-2 fuzzy weight adjustment for back propagation neural networks with application in time series prediction. *Elsevier Information Sciences, 260,* 1–14.
20. Song, Q., & Chissom, B. S. (1993). Fuzzy time series and its model. *Fuzzy Sets Systems, 54* (3), 269–277.
21. Song, Q., & Chissom, B. S. (1993). Forecasting enrollments with fuzzy time series—part I. *Fuzzy Sets Systems, 54*(1), 1–9.
22. Song, Q., & Chissom, B. S. (1994). Forecasting enrollments with fuzzy time series—part II. *Fuzzy Sets Systems, 62*(1), 1–8.
23. Yu, H. K. (2005). Weighted fuzzy time series models for TAIEX forecasting. *Physica A, 349,* 609–624.
24. Chen, Mu-Yen. (2014). A high-order fuzzy time series forecasting model for internet stock trading. *Elsevier, Future Generation Computer Systems, 37,* 461–467.
25. Huarng, K., Yu, H. K., & Hsu, Y. W. (2007). A multivariate heuristic model for fuzzy time-series forecasting. *IEEE Transactions on Systems Man and Cybernetics, Part B Cybernetics, 37*(4), 836–846.
26. Chen, S. M., Chu, H. P., & Sheu, T. W. (2012). TAIEX forecasting using fuzzy time series and automatically generated weights of multiple factors. *IEEE Transactions Systems, Man, Cybernetics, Part A: Systems and Humans, 42*(6), 1485–1495.
27. Chen, S. M., & Chen, C. D. (2011). TAIEX forecasting based on fuzzy time series and fuzzy variation groups. *IEEE Transactions on Fuzzy Systems, 19*(1), 1–12.
28. Chen, S. M., & Kao, P. Y. (2013). TAIEX forecasting based on fuzzy time series, particle swarm optimization techniques and support vector machines. *Information Science, 247,* 62–71.
29. Cai, Q., Zhang, D., Zheng, W., & Leung, S. C. H. (2015). A new fuzzy time series forecasting model combined with ant colony optimization and auto-regression. *Knowledge-Based Systems, 74,* 61–68.
30. Chen, S. M. (1996). Forecasting enrollments based on fuzzy time series. *Fuzzy Sets Systems, 81*(3), 311–319.
31. Yu, T. H. K., & Huarng, K. H. (2008). A bivariate fuzzy time series model to forecast the TAIEX. *Expert Systems with Applications, 34*(4), 2945–2952.
32. Yu, T. H. K., & Huarng, K. H. (2010). Corrigendum to "A bivariate fuzzy time series model to forecast the TAIEX. *Expert Systems with Applications, 37*(7), 5529.
33. Huarng, K., & Yu, T. H. K. (2006). The application of neural networks to forecast fuzzy time series. *Physica A, 363*(2), 481–491.
34. Karnik, N. N., & Mendel, J. M. (1999). Applications of type-2 fuzzy logic systems to forecasting of time-series. *Elsevier, Information Sciences, 120,* 89–111.
35. Elanayar, V. T. S., & Shin, Y. C. (1994). Radial basis function neural network for approximation and estimation of nonlinear stochastic dynamic systems. *IEEE Transactions on Neural Networks, 5*(4), 594–603.
36. Park, Y. R., Murray, T. J., & Chen, C. (1996). Predicting sun spots using a layered perceptron neural network. *IEEE Transactions on Neural Networks, 1*(2), 501–505.
37. TAIEX. [Online]. Available: http://www.twse.com.tw/en/products/indices/tsec/taiex.php
38. Ma, Q., Zheng, Q., Peng, H., Zhong, T., & Xu, L. (2007) Chaotic time series prediction based on evolving recurrent neural networks. In *Proceedings of the Sixth International Conference on Machine Learning and Cybernetics, Hong Kong*, 2007.
39. Koskela, T., Lehtokangas, M., Saarinen, J., & Kaski, K. (1996). Time series prediction with multilayer perceptron, FIR and Elman neural networks. In *Proceedings of the World Congress on Neural Networks* (pp. 491–496).

# Chapter 6
# Conclusions

This is the concluding chapter of the book. It self-reviews the book and examines the possible scope of future research directions in the research arena covering time-series forecasting.

## 6.1 Conclusions

The book deals with two distinct problems of time-series prediction. The first problem refers to predicting the next time-point value of a time-series from its current and preceding values. It primarily deals with uncertainty management in the prediction process using fuzzy and neural techniques. Several models of fuzzy and neural techniques are available in the literature. However, our present concern is to deal with uncertainty within a partition and across partitions of a time-series. The within partition uncertainty is represented by a set of Gaussian type-1 membership functions (MFs), which together is represented by an interval type-2 fuzzy set. The time-series lying across partitions are represented by distinct interval type-2 fuzzy sets. In Chap. 2, we solved the uncertainty management problem in time-series prediction by using prediction rules with single antecedent and single consequent, and secondary factor is used to select the right rule for fuzzy reasoning and prediction.

Like Chaps. 2, 3 also deals with prediction of the next time-point in a time-series, where the secondary factor is used as part of the antecedent in the prediction rule. Such formulation reduces the complexity to select the firing rule induced by secondary factor, however at the cost of additional reasoning complexity. In addition, the chapter demonstrated special situations, when there exists only one contiguous region of data points in a partition rather than bunches of regions. The single region of contiguous data points in a partition has been modeled



A. Konar and D. Bhattacharya, *Time-Series Prediction and Applications*,
Intelligent Systems Reference Library 127,
DOI 10.1007/978-3-319-54597-4_6

with type-1fuzzy sets. Thus in Chap. 3 we introduce a mixed type-1/type-2 reasoning in antecedent/consequent of the prediction rules. An automatic approach to adaptive tuning of membership functions is undertaken here to imbibe the prediction results with the current trend.

Chapters 4 and 5 are concerned with learning a time-series moves by two alternative approaches. In Chap. 4, we propose a clustering based approach for learning time-series moves, while in Chap. 5 we learn the moves by a number of neural networks, each engaged in prediction of the time-series by concurrently activating all possible rules having the common antecedent. Both the techniques presented in Chaps. 4 and 5 are novel in the context of time-series prediction. The novelty in Chap. 4 lies in all the three basic steps, including segmentation, clustering and knowledge representation using dynamic stochastic automaton. On the other hand, novelty in Chap. 5 primarily lies in structural organization of the neural network to fire multiple rules with common antecedent simultaneously.

All the four major chapters covered in the book are complete in themselves, covering problem description, analysis, and methodology and performance analysis with existing works. Parameter variations and system validation has also been undertaken in Chaps. 2 and 3.

## 6.2 Future Research Directions

The study undertaken in the book is primarily concerned with prediction of economic time-series. The theory can be extended for other time-series as well. For example, the segmentation algorithm developed in Chap. 4 is relevant for the economic time-series only, as it considers rise/fall and zero slopes only to identify fewer structures for a selected economic time-series, which of course varies for a different economic time-series. These structures are semantically important for economic time-series only. To explore structures hidden in other time-series, the segmentation algorithm should be equipped with relevant information of the subjective domain of the series concerned. For instance, to segment electroencephalograph (EEG) time series, we need to extract certain short duration signals, representing specific brain activity, such as motor imagery about left or right hand, dreaming, smart movement related planning and error detection in one's activity by himself. Each of these activities has special wave shapes, segmentation of which is very relevant for an EEG signal. The clustering algorithm used in Chap. 4 can be replaced and or augmented by more sophisticated means to improve qualitative clustering and computational overhead.

In addition, the interval type-2 fuzzy set based reasoning used can be replaced by more sophisticated General type-2 fuzzy Set induced reasoning. The uncertainty involved within and across partitions here can be represented by General Type-2 Fuzzy Sets, which would help better reasoning results and thus better prediction.

# Index




A. Konar and D. Bhattacharya, *Time-Series Prediction and Applications*,
Intelligent Systems Reference Library 127,
DOI 10.1007/978-3-319-54597-4

Printed in the United States
By Bookmasters